Physics for Anesthesiologists and Intensivists

Antonio Pisano

Physics for Anesthesiologists and Intensivists

From Daily Life to Clinical Practice

Second Edition

Springer

Antonio Pisano
Cardiac Anesthesia and ICU
AORN Dei Colli, Monaldi Hospital
Naples
Italy

ISBN 978-3-030-72046-9 ISBN 978-3-030-72047-6 (eBook)
https://doi.org/10.1007/978-3-030-72047-6

This Springer imprint is published by the registered company Springer Nature Switzerland AG
The registered company address is: Gewerbestrasse 11, 6330 Cham, Switzerland

To my wife Marcella and my son Matteo…
… And to all brains that are never satiated with knowledge

Preface

The idea of a second edition of *Physics for Anesthesiologists* was born almost immediately after its release, thinking of all the other possible topics that had not come to my mind or that I had decided to leave out, and it has been consolidated also thanks to the many appreciations that the book has received from colleagues from all over the world in the four years that have passed since then, including those contained in a review by professor Spieth (whom I take this opportunity to thank), who fully grasped and perfectly synthesized the "soul" of *Physics for Anesthesiologists* (Spieth PM. Physics for Anesthesiologists: From Daily Life to the Operating Room. Anesth Analg 2018; 127(2):e19).

This soul remains intact and is perhaps further exalted, in this second edition, that I decided to title *Physics for Anesthesiologists and Intensivists* just to remark that all topics discussed are of close interest to both specialties (this could have been less evident from the title of the first edition, since, unlike what happens in my country, in many other parts of the world the two careers are completely independent).

I hope readers have fun reading this book as much as I enjoyed writing it. The features of the first edition all remain in this new book: the physical laws explained in detail, often through their mathematical derivation, and not only mentioned; their implications (or applications) in clinical practice clearly highlighted and discussed; the often informal and blunt language; the many examples both from daily life and from the clinical settings in which we all work; the introductory boxes "Where's Physics," which anticipate the contents of each chapter at a glance; the curiosities and insights.

However, there is also much more. First of all, more physics. In addition to all topics covered in the previous edition, new chapters deal with notions of electricity, geometrical optics, waves, and harmonic motion which are relevant to the clinical practice of anesthesia and critical care. Moreover, the discussion about many physical laws already dealt with in the first edition has been expanded and enriched, and several hints on "modern physics" (e.g., quantum mechanics) have been added. Second, more clinical implications and applications: extracorporeal membrane oxygenation, the pathophysiology of resorption atelectasis and systolic anterior motion, temperature homeostasis and monitoring, optical fibers, and defibrillators are only some examples. Furthermore, important aspects of invasive pressure monitoring and ultrasonography, some of which were only briefly hinted in the previous edition, are now deeply discussed. Also, more examples from everyday life (we will even talk about, among other things, popcorn, roofs blown away by the wind, and why the setting sun has, actually, already set), more figures, and more curiosities. Finally, I added an initial chapter on, practically, all the mathematics which we will need throughout the book, which also includes an enjoyable (I hope) digression on how geometry and trigonometry may help us during the ultrasound-guided positioning of a central venous catheter. However, you can also decide to skip it and consult it only if you have mathematical troubles while reading the various chapters.

This time I have to thank more people. In addition to my wife Marcella and my son Matteo for patiently putting up with me while I wrote hours and hours on my laptop, I thank, again, Dr. Viviana Carillo, medical physicist and one of my best friends, and Prof. Enrico Borriello, who is a theoretical physicist at the Arizona State University in Tempe, AZ, for their advices and their support for the physics part. As for the clinical aspects, I thank Dr. Concetta Palmieri, another dear friend and a pediatric anesthetist in Naples, Italy, for her opinion on the section about trigonometry and central venous catheterization (I was afraid, in fact, of having exaggerated, but she reassured me; in case, you know who to blame!). Finally, I thank Dr. Rasul Ashurov, cardiac surgery resident, who is the author of the illustration in Figure 13.5; Dr. Vittorio Palmieri, a cardiologist at my hospital with whom I have sometimes the pleasure of having interesting conversations, for having promptly provided me with some echocardiographic images (which, however, I no longer needed to include in the book); all the kind and very professional people who work for Springer-Nature, including my editor Andrea Ridolfi, who gave me valuable advice; and all the colleagues who have shown their appreciation for the first edition and those who will want to grant me the honor of reading this new book.

Naples, Italy Antonio Pisano

Preface to the First Edition

Physics is everywhere. This is not surprising, really, given that its field of interest is the description of how the matter around us behaves, from the "little" world of atoms and subatomic particles to the entire Universe, passing for the objects and situations of daily life. Moreover, it is not a mystery that there are specific laws of physics at the basis of a lot of things we everyday do as anesthesiologists; most university training courses in anesthesiology, indeed, include the study of physics, and it is not uncommon to find more or less complex physics equations strewn among the pages of landmark anesthesia textbooks. These equations, however, are often skipped or soon forgotten.

Unfortunately, in fact, most people consider physics as an abstract and difficult matter (sometimes incomprehensible), if not even of little practical use. Conversely, there is no modern technological device (including smartphones) which does not rely on some conquest of physics. Furthermore, if you are an inquisitive person, physics can answer many interesting questions, sometimes in a surprisingly clear and illuminating way. For example, the ideal gas law tells us how to make a good coffee (with the "moka" pot); Henry's law explains the behavior of carbonated drinks and Champagne corks; Laplace's law reveals some secrets of bubbles; Bernoulli's theorem teaches us how an airplane can fly; thermodynamics shows the direction in which time flows; the study of heat and state changes of matter has many implications in everyday life, from our kitchens to the building of bridges. Moreover, there is a law of physics which accounts for the blue color of oceans, and a different one which explains why also the sky is blue; also, there is the physics that allows you to enjoy a concert, bats to avoid obstacles, and police to make you a speeding fine.

If, apart from being an inquisitive person, you are an anesthesiologist or an intensive care practitioner, you might take advantage (and, probably, some fun) in discovering, or rediscovering, the important implications (and applications) of physics in your daily clinical practice; for example, the abovementioned laws have something to do, approximately in the same order, with oxygen cylinders, blood gas analysis, airway management, hemodynamic monitoring, anesthetic vaporizers, pulse oximeters, near-infrared spectroscopy, and ultrasounds. And there is much more.

This book is intended for all who are interested in anesthesiology and intensive care medicine, from medical students to experienced clinicians, and can be read at different levels: to fully understand, in the training stage, the basic science that lies behind many aspects of anesthesia and intensive care medicine, and which is often only hinted elsewhere; to have a higher awareness about why we do what we do, as anesthesiologists, how the devices we commonly use work, and why we should not always blindly trust them; finally, just for curiosity, culture, or fun (maybe skipping some equation). Although of varying difficulty, all topics are discussed through many examples from daily life and are accompanied by a lot of color illustrations for extra clarity.

Moreover, the topic selection reflects both the interests and the basic mathematical knowledge of a clinician (such as the author himself is); accordingly, each aspect of physics discussed is strictly related to the clinical practice of anesthesia (and/or intensive care medicine), can be easily understood according to the recollections of high school (often refreshed within the text or in practical "boxes"), and allows to make interesting and enlightening comparisons with everyday life.

I would like to thank my wife Marcella and my son Matteo, to whom this book is dedicated, for their patience during the exciting but heavy commitment of writing a book. I would also like to thank Dr. Viviana Carillo and Dr. Pietro Castellone, two brilliant medical physicists (and friends), for their advices and support.

Naples, Italy
2017

Antonio Pisano

Contents

Part I

Before Starting

A Little Math we Will Need Throughout the Book and a Clinical Application Immediately

1

Contents

1.1 We Must "Speak Mathematics" (But Don't Worry About It)

The physicist Norman Packard said that "a physicist is a mathematician with a feeling for reality" [1]. Beyond the humorous connotation of this statement (at least as I understand it), it is a matter of fact that physics and mathematics are inextricably linked. Regardless of the reasons, still partly mysterious, why the mathematical language is so incredibly effective in describing the "physical world" [2, 3], even in this book we cannot do without a little mathematics to fully understand the physics that lies behind our "clinical" world, namely that of anesthesiology and intensive care medicine. However, you can rest assured: all physical laws discussed in the next 20 chapters can be dealt with in an understandable and, at the same time, sufficiently rigorous way using the math being taught in all high schools in the world.

Of course, there is always the possibility to skip proofs and mathematical derivations and get straight to the point (as I myself will sometimes suggest to the less fond of numbers and equations), but I think there is much more pleasure in

A. Pisano, *Physics for Anesthesiologists and Intensivists*, https://doi.org/10.1007/978-3-030-72047-6_1

understanding how you get to certain formulas and laws. After all, it will be sufficient to rely on some algebra, a bit of geometry, a little trigonometry, and just a couple of notions of calculus (very rarely). However, none of this will ever be taken for granted: even if there is something you don't remember, the mathematical rules and formulas according to which each step is carried out will always be at your fingertips in each chapter. Moreover, if you really believe your math is very rusty, you will be ready to face the whole book serenely after reading the few pages that follow. In this chapter, we will also take the opportunity to immediately make some considerations relating to our daily clinical practice.

1.2 Playing with Equations

If I have not counted wrong (which would be rather embarrassing in a chapter on mathematics!), there are 265 equations in this book (leaving out those in this chapter). Sometimes, they are only starting points or intermediate steps which need to be rearranged, substituted according to other equations, simplified, or resolved for a certain variable to lead us, eventually, to the desired result. If you think it might be helpful to brush up on your algebra skills, we can try solving together the three following riddles (or you can try to do it by yourself, just for fun, after covering the part with the equations).

A few years ago, on a summer vacation at a beach resort, an entertainer approached me and asked: "a brick weighs one kilogram plus half a brick: how many kilograms does it weigh?." I began to wave the index finger of my right hand in the air with an absorbed gaze, and after a few seconds I gave her the right answer. I had solved the following equation in my head:

$$x = 1 + \frac{x}{2} \tag{1.1}$$

where x is the brick's weight in kilograms (to be precise, as discussed in Box 3.1, and in Chap. 14, what we commonly call the "weight" of an object is, actually, its *mass*).

To solve Eq. (1.1), we can start by multiplying both its left-hand side and its right-hand side by 2:

$$2x = 2 + \frac{2x}{2} \tag{1.2}$$

that is:

$$2x = 2 + x \tag{1.3}$$

which becomes:

$$2x - x = 2 \tag{1.4}$$

and, finally:

$$x = 2 \tag{1.5}$$

This was easy. Let's try the next one.

We have a box containing 1 kg chocolates and one containing 1 kg candies. We only know that candies are 100 more than chocolates and that a chocolate weighs three times the weight of a candy. How many chocolates and candies do we have? (It seems to be back in school, I know: it's a nightmare!)

We can start by calling x the number of chocolates and y that of candies. Evidently, it is:

$$y - x = 100 \tag{1.6}$$

If x chocolates weigh 1 kg, then one chocolate weighs (in kg) $1/x$. Similarly, one candy weighs $1/y$. Since we said that a chocolate weighs three times the weight of a candy, we can write:

$$\frac{1}{x} = 3\frac{1}{y} \tag{1.7}$$

We can invert the numerator and denominator at both sides without altering the "meaning" of the equation, which becomes:

$$x = \frac{y}{3} \tag{1.8}$$

or

$$y = 3x \tag{1.9}$$

After substituting for y according to Eq. (1.9), Eq. (1.6) becomes:

$$3x - x = 100 \tag{1.10}$$

that is,

$$2x = 100 \tag{1.11}$$

or

$$x = 50 \tag{1.12}$$

Finally, according to Eq. (1.9), it is:

$$y = 150 \tag{1.13}$$

That wasn't difficult either, was it? Maybe the third and last riddle will be just a bit harder. I got it from the "Italian Committee for the Investigation of Claims of Pseudosciences" (CICAP) website [4].

An elderly lady goes from house to house to sell the apples she has in her basket. At the first house, she sells half of the apples contained in the basked plus half an apple; at the second house, she sells half of the apples left in the basket plus half an apple; at the third house, she again sells half of the apples left in the basket plus half an apple, and so she sold all the apples she had in the basket. How many apples did the lady have in the beginning?

If we call x the initial number of apples contained in the basket, it must be:

$$x-\left(\frac{x}{2}+\frac{1}{2}\right)-\left(\frac{x-\left(\frac{x}{2}+\frac{1}{2}\right)}{2}+\frac{1}{2}\right)-1=0 \tag{1.14}$$

where $x/2 + 1/2$ is the number of apples sold at the first house; $[x-(x/2 + 1/2)]/2 + 1/2$ is the number of apples sold at the second house (since $x-(x/2 + 1/2)$ is the number of apples left in the basket after the first sale); and 1 is the apple sold at the third house (in fact, if there are no more apples after selling half of the apples left in the basket plus half an apple, this can only mean that only one apple was left after the second sale).

How can we solve Eq. (1.14)? As for Eq. (1.1), the most convenient thing to do is to multiply both sides by 2 (we can do whatever we want, as long as we do it for both sides):

$$2x-\frac{2x}{2}-\frac{2}{2}-\frac{2\left(x-\left(\frac{x}{2}+\frac{1}{2}\right)\right)}{2}-\frac{2}{2}-2=2*0=0 \tag{1.15}$$

Now we can simplify all the fractions (which is equivalent to dividing the numerator and denominator by the same number):

$$2x-x-1-\left(x-\left(\frac{x}{2}+\frac{1}{2}\right)\right)-1-2=0 \tag{1.16}$$

Once again, we can multiply all by 2:

$$4x-2x-2-2x+x+1-2-4=0 \tag{1.17}$$

and, after some addition and subtraction:

$$x-7=0 \text{ or } x=7 \tag{1.18}$$

This was just a little warm-up to prepare you for the equations contained in the next chapters. In some of them, you will find a few different operators such as square roots, exponentials, logarithms, trigonometric functions, and, occasionally, derivatives or integrals (see below). However, we will never have to do more complex calculations than those carried out in the examples above.

1.3 A Bit of Geometry and Trigonometry

Every so often throughout this book, we will have to remember some simple formula, rule, or theorem of geometry. For example, in order to define *surface tension* in Chap. 6, we will need the formula for the area of a rectangle (but I think we all remember that it is base x height!). In the same chapter, we will use the formula for the surface area of a sphere to derive *Laplace's law* and then discover what this law has to do with: tracheal tube cuffs; the role of the reservoir bag in anesthesia breathing systems; air embolism; dilated cardiomyopathy; aortic aneurysms; alveolar collapse; and catheter obstruction.

Given a sphere with radius r (e.g., the basketball in Fig. 1.1a), its surface area S is:

$$S = 4\pi r^2 \tag{1.19}$$

where π is the ratio of the circumference to the diameter of any circle. We will find this important constant of nature in many equations in the next chapters. It is an *irrational number*, that is a number which cannot be expressed as the ratio of two integer numbers (hence the name "irrational") since it has an *infinite* series of decimals which does not terminate with a repeated sequence [5]. In other words, π is a number that cannot be defined *exactly*. However, for most calculation purposes it is usually approximated to 22/7 or 3.14.

In Chap. 17, Eq. (1.19) will also allow us to understand the so-called *inverse square law*, which tells us how far to stay away from a source of ionizing radiation (e.g., during a chest X-ray in the intensive care unit).

A geometric figure that will recur quite often in the book is the cylinder. In Chap. 8, we will discuss the *Hagen–Poiseuille equation*, which describes the flow of a fluid (i.e., a liquid or a gas) through a cylindric conduit and will allow us to make interesting considerations about endotracheal or tracheotomy tubes, central venous catheters, epidural catheters, and hemodynamics (e.g., we will derive the well-known relationship among mean arterial pressure, central venous pressure, cardiac output, and systemic vascular resistance from Hagen–Poiseuille equation). Moreover, in order to derive several physical laws and equations of great interest for anesthesiologists and intensivists, we will need to remember that the *volume* V of a cylinder with base area A, radius r, and height (or length) h (see Fig. 1.1b) is:

$$V = Ah = \pi r^2 h \tag{1.20}$$

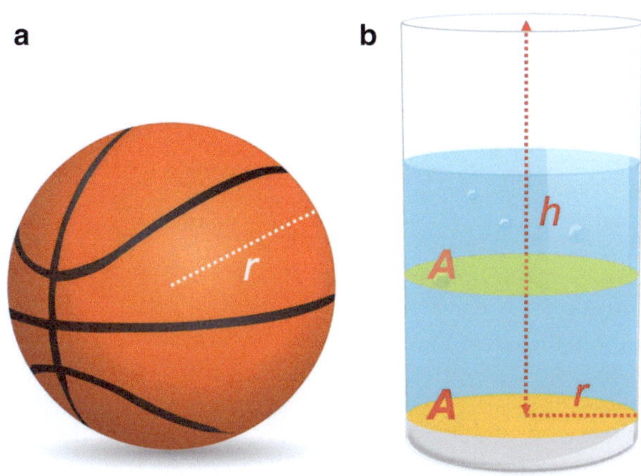

Fig. 1.1 (**a**) The surface area of a sphere with radius r is $4\pi r^2$. (**b**) The volume of a cylinder with base (or cross-sectional) area A and height (or length) h is Ah, with A being the area of a circle with radius r, that is πr^2

where πr^2 is the area of a circle with radius r, that is the base (or the cross-sectional) area of the cylinder.

We will have to keep in mind Eq. (1.20) when we derive the formula for the *work of breathing* (WOB) from the physical definition of mechanical work (Chap. 8); when we prove *Stevin's law*, which has important implications for transducer leveling during invasive pressure monitoring (Chap. 9); when we derive the formula for the natural frequency of an invasive pressure monitoring system from the equations of *simple harmonic motion* (Chap. 19); and when we see how stroke volume (and, hence, cardiac output) is measured using Doppler analysis (Chap. 20).

1.3.1 Angles and Triangles

As Leonard Susskind and George Hrabovsky remember in their beautiful book [6], trigonometry (i.e., that part of mathematics which studies the relationship between angles and side lengths of triangles) is everywhere in physics! Not surprisingly, therefore, the sine or cosine (or some other *trigonometric function*) of a certain angle will appear quite often in the equations and physical laws we will discuss in this book. For example, whenever there are *vectors* at stake (Chapters 6, 7, 13, 14, and 20) and, in particular, when we will talk about topics such as: ramps, friction, and Trendelenburg position (Chap. 14); the scattering of electromagnetic waves and its relation with either near-infrared spectroscopy (NIRS) oximetry or protection from ionizing radiations (Chap. 17); the laws of light reflection and refraction and their application in clinical practice (Chap. 18); and the Doppler effect in cardiac ultrasonography and minimally invasive cardiac output monitoring (Chap. 20). Moreover, *sinusoidal waves* (see Sect. 1.4) will be fundamental when we discuss light and pulse oximetry (Chap. 16), as well as mechanical oscillations, resonance, and their implications for invasive pressure monitoring systems (Chap. 19).

However, all the trigonometry we will need can be brought to mind with a quick glance at Fig. 1.2, which shows a right triangle drawn inside a circle whose center coincides with the origin of a *rectangular* (or *Cartesian*) *coordinate system.*

In a *right triangle*, we define the sine (sin) of an angle ϑ as the ratio of the length of the side which is *opposite* to ϑ to that of the hypotenuse, and the cosine (cos) of ϑ as the ratio of the length of the side which is *adjacent* to ϑ to that of the hypotenuse. That is, in Fig. 1.2:

$$\sin\vartheta = \frac{a}{h} \tag{1.21}$$

and

$$\cos\vartheta = \frac{b}{h} \tag{1.22}$$

where a, b, and h are, respectively, the length of the side opposite to ϑ, the length of the side adjacent to ϑ, and the length of the hypotenuse (i.e., the side opposite to the right angle).

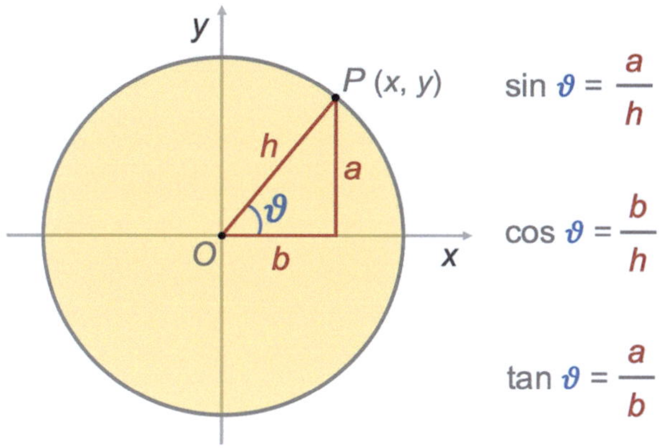

Fig. 1.2 The definitions of sine (sin), cosine (cos), and tangent (tan) of an angle ϑ in the context of a right triangle drawn inside a circle whose center coincides with the origin of a Cartesian coordinate system

Equations (1.21) and (1.22) allow us to express the position (i.e., the *x* and *y* coordinates in a Cartesian coordinate system) of any point *P* according to the length of the segment connecting the origin *O* of the Cartesian coordinate system to the point *P*, and to the width of the angle which this segment makes with one of the two axes [6]. As it is easy to see, this segment coincides with the hypotenuse *h* in Fig. 1.2. Accordingly, with respect to the angle ϑ made by the segment *OP* (= *h*) with the *x*-axis, it is:

$$x = b = h\cos\vartheta \tag{1.23}$$

and

$$y = a = h\sin\vartheta \tag{1.24}$$

where Eqs. (1.23) and (1.24) are none other than Eqs. (1.22) and (1.21), respectively. As we will see throughout the book, equivalences like these are often used in physics: for example, as discussed in Chap. 13, to find the *component* of a vector (i.e., a physical quantity which needs a *direction* in addiction to a "magnitude" to be defined) along any direction.

Table 1.1 shows the values of the sine and cosine of some angles. Let's just focus on the first four of them (0°, 30°, 45°, and 90°). Having another look at Fig. 1.2, you can easily realize why $\sin 0° = \cos 90° = 0$ and $\cos 0° = \sin 90° = 1$: in fact, when $\vartheta = 0°$, the side *a* "disappears" and, hence, it is $a/h = 0$, while the side *b* becomes equal to the hypotenuse, and so it is $b/h = 1$; conversely, when $\vartheta = 90°$, it is $a = h$ and $b = 0$: accordingly, it is $a/h = 1$ and $b/h = 0$. It should also be clear why $\sin 45° = \cos 45°$; in fact, for $\vartheta = 45°$, our triangle becomes an *isosceles right triangle*, that is a right triangle with two equal angles (since, as mentioned in Box 1.1, the sum of the three angles of any triangle is always 180°) and *two equal sides*; accordingly, it is $a = b$ and, hence, $a/h = b/h$. Going on with a similar reasoning, it is not even difficult to understand why the values of the sine and cosine of these angles are between 0 and 1 (in fact, the sides *a* and *b* are never longer than the hypotenuse) and why $\sin\vartheta$ increases, while $\cos\vartheta$ decreases, as the width of ϑ increases from 0° to 90°.

Table 1.1 The values of the sine (sin) and cosine (cos) of some angles (approximated to the nearest hundredth)

ϑ (°)	sin ϑ	cos ϑ
0	0	1
30	0.5	0.87
45	0.71	0.71
90	1	0
120	0.87	−0.5
180	0	−1
270	−1	0
360	0	1

All trigonometric functions are generalized to angles of any width (here we don't care how), and this is why angles greater than 90° (and *negative* values of their sine and cosine) also appear in Table 1.1 although there are certainly no angles >90° in a right triangle. Ultimately, as we will see better in Sect. 1.4, the sine and cosine of any angle only assume values between −1 and 1, which increase with the angle width in some width intervals and decrease with it in others.

Box 1.1: Little Memories on Angles and Triangles

Triangle postulate

The sum of all the angles of a triangle is two right angles (i.e., 180°).

Triangle inequality theorem

The sum of the lengths of any two sides of a triangle is always greater than the length of the third side.

Complementary angles

Two angles are said to be complementary when their sum is 90° (e.g., 30° and 60° are complementary to each other).

As you may remember, sine and cosine are not the only trigonometric functions (there are four more "basic" ones). However, the only other trigonometric function we will encounter in the book (see below and Chap. 14) is the *tangent* (tan), defined as the ratio of the sine to the cosine of an angle. In Fig. 1.2, hence, it is:

$$\tan\vartheta = \frac{\sin\vartheta}{\cos\vartheta} = \frac{ah}{hb} = \frac{a}{b} \tag{1.25}$$

That is, in a right triangle, the tangent of an angle ϑ is the ratio of the length of the side which is opposite to ϑ to the length of the side which is adjacent to ϑ (we have simply substituted for sin ϑ and cos ϑ with a/h and b/h, respectively, according to Eqs. (1.21) and (1.22)).

As we will discuss later, remembering the definition of tangent of an angle is important to discuss notions such as velocity (see below) and gradient (Chap. 4), and may also have to deal with ultrasound-guided central venous catheter insertion (Sect. 1.3.2) or with the risk of patient slide in the Trendelenburg position (Chap. 14).

Finally, all trigonometric functions have their *inverse function*. For example, the arcsine (arcsin or $\sin^{-1}$) is the inverse function of the sine. That is, if it is:

$$\sin\vartheta = x \tag{1.26}$$

then it is:

$$\arcsin(x) = \vartheta \tag{1.27}$$

Just remember this when we discuss geometrical optics and, in particular, optical fibers (Chap. 18).

Something else we will need to keep in mind about angles and triangles is summarized in Box 1.1.

1.3.2 Pythagorean Theorem, Trigonometry, and Central Venous Catheters

In general, there is the following relationship between the sine and cosine of any angle:

$$\sin^2\vartheta + \cos^2\vartheta = 1 \tag{1.28}$$

If ϑ is one of the two angles other than 90° of a right triangle (as in Fig. 1.2), then Eq. (1.28) becomes something you certainly remember. In fact, after substituting for $\sin\vartheta$ and $\cos\vartheta$ according to Eqs. (1.21) and (1.22), respectively, Eq. (1.28) can be written as:

$$\left(\frac{a}{h}\right)^2 + \left(\frac{b}{h}\right)^2 = 1 \tag{1.29}$$

that is:

$$\frac{a^2}{h^2} + \frac{b^2}{h^2} = 1 \tag{1.30}$$

or, after multiplying all by h^2:

$$a^2 + b^2 = h^2 \tag{1.31}$$

Now you surely recognize it: Eq. (1.28) is none other than the "trigonometric version" of the *Pythagorean theorem* [6], which states that, *in a right triangle, the square of the hypotenuse is equal to the sum of the squares of the other two sides* (see Fig. 1.3).

As will be mentioned in Chap. 13, the Pythagorean theorem can be used to find the sum of two vectors whose directions are perpendicular to each other: for example, the vector sum (or *resultant*) of two forces acting perpendicular to each other is

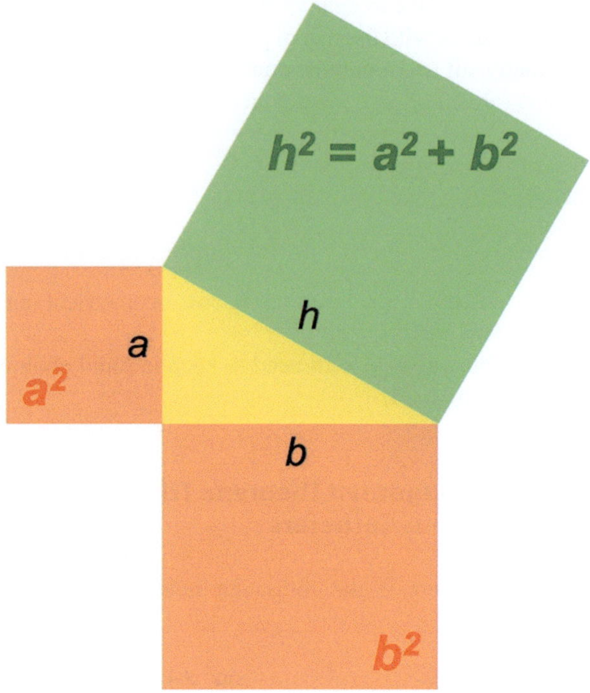

Fig. 1.3 The Pythagorean theorem states that, in a right triangle, the square of the hypotenuse *h* (i.e., the side opposite to the right angle) is equal to the sum of the squares of the other two sides (*a* and *b*)

the hypotenuse of a right triangle whose side lengths are those of the vectors representing the two forces (all will be clearer in Chap. 13). Moreover, in Chap. 21, the Pythagorean theorem will help us to derive, even, the Einstein's special relativity equation for *time dilation*, with which we will (nicely) make fun of the surgeons.

And here we are at the "clinical application" promised in the title of this chapter. As proposed by Piton and colleagues [7], the Pythagorean theorem may also provide some help during ultrasound-guided placement of a (central) venous catheter, especially when teaching the technique to young fellows and, in general, to physicians with poor experience in ultrasound-guided vascular access.

If you place the ultrasound probe perpendicular to the skin just above the vein and insert the needle making an angle of 45° with the skin at a distance x from the center of the probe equal to the depth d of the vein under the probe, you ideally drew an isosceles right triangle (see above) whose hypotenuse represents the length L by which the needle must be introduced through the skin to reach the vein (Fig. 1.4). This length can therefore be calculated according to the Pythagorean theorem:

$$L^2 = x^2 + d^2 \tag{1.32}$$

If, as stated, you choose a distance $x = d$, we can replace x with d in Eq. (1.32), which becomes:

$$L^2 = d^2 + d^2 = 2d^2 \tag{1.33}$$

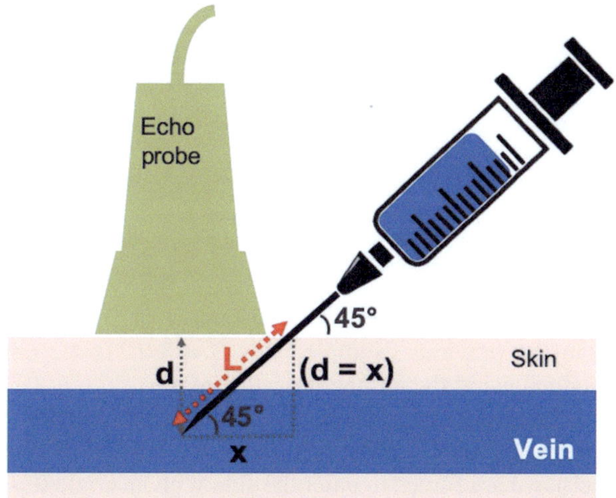

Fig. 1.4 A Pythagorean (or also a trigonometric) approach to central venous catheterization (see text). Modified with permission from Piton et al. [7] (BioMed Central Ltd.)

or

$$L = \sqrt{2d^2} = d\sqrt{2} \tag{1.34}$$

where $\sqrt{2}$ is another irrational number (like π), whose value is 1.414213562 followed by an infinite number of digits which do not repeat regularly [5]; however, we can approximate it to 1.4.

Hence, if the vein is located at a depth d of, say, 3 cm (think, for example, of a femoral vein in an obese patient), the needle must be introduced by roughly $3 \times 1.4 = 4.2$ cm. Following this procedure allows you to assess if the needle length is sufficient, to know in advance "when" you will enter the vein (i.e., how far you need to introduce the needle) and, if the vein is parallel to the skin surface, to enter the vein itself with an angle of 45°, thus minimizing the risk of both transfixion and guidewire kinking during dilation [7].

We could come to the same conclusions also using trigonometry. For example, remembering that the tangent of an angle is the ratio of the length of the opposite side to that of the adjacent side (Eq. 1.25), and that it is tan 45° = 1, in Fig. 1.4 it must be:

$$\frac{d}{x} = \tan 45^\circ = 1 \tag{1.35}$$

or

$$d = x \tag{1.36}$$

That is, as stated above, if you want to enter the skin with an angle of 45°, you must choose a point whose distance x from the probe is equal to the depth d of the vein under the probe in order to enter the vein just below the center of the probe [8].

Furthermore, we can calculate the length L by which the needle must be introduced through the skin to reach the vein also remembering that it is:

$$\sin 45^{\circ} = \frac{d}{L} \tag{1.37}$$

or

$$L = \frac{d}{\sin 45^{\circ}} \tag{1.38}$$

Assuming sin 45° = 0.71 (see Table 1.1), for a vein located at a depth $d = 3$ cm as in the previous example, it is $L = 3/0.71 \cong 4.2$ cm (which is the same result obtained using the Pythagorean theorem).

Of course, if there is no space to make the puncture at a distance from the probe equal to the depth of the vein (as in the case of an obese patient with a very short neck), we are forced to enter the skin with an angle $\vartheta > 45°$ if we want to see (sonographically) the needle entering the vein, especially if the long-axis (or "in-plane") technique is not possible. Using trigonometry, one could even calculate exactly the angle with which to enter the skin. Imagine to make the puncture at a distance x_1 from the probe which is lower than the depth d. Again (similarly to what we said for Fig. 1.4, but with the difference that $x = d$ is replaced by $x_1 < d$ and that the angle is not 45° but an unknown angle ϑ), to see the needle entering the vein just below the probe, it must be:

$$\frac{d}{x_1} = \tan \vartheta \tag{1.39}$$

Accordingly, the angle ϑ can be calculated as:

$$\vartheta = \tan^{-1}\left(\frac{d}{x_1}\right) \tag{1.40}$$

where $\tan^{-1}$ stands for arctangent (arctan), i.e., the inverse of the tangent (just like the arcsine is the inverse of sine).

Let's take an example. If the vein is located at a depth $d = 3$ cm, but I can only make the puncture at a distance x_1 of 2 cm from the probe, the angle of needle insertion should be:

$$\vartheta = \tan^{-1}\left(\frac{3}{2}\right) = \tan^{-1}(1.5) \cong 56^{\circ} \tag{1.41}$$

Obviously, I challenge anyone to really measure distances and angles exactly or to calculate trigonometric functions while inserting a central venous catheter. However, I hope all this can give an underlying "mathematical sense" to what we do in our daily clinical practice.

1.4 One Last Little Effort: Functions, Limits, Derivatives, and Integrals

Just a few words. In this book, I avoided any topic which needed skills of differential or integral calculus. Accordingly, here we will only make some consideration about functions and brush up on a few definitions such as that of derivative and integral: these notions will occasionally appear in some chapters and will help us better understand, or describe more precisely, what we are discussing.

As you will certainly remember, leaving out rigorous mathematical definitions, a *function* is a "law" which regulates the interdependence of variable quantities [9]. In other words, when a certain variable y (referred to as the "dependent variable") assumes a single specific value for each value assumed by another variable x (called the "independent variable")," we say that y is a function of x, and we indicate this with the following notation:

$$y = f(x) \tag{1.42}$$

As we all know, a function can be usually represented by a *graph*, i.e., a curve made up by all the points identified in a Cartesian coordinate system by the coordinates x, $f(x)$ for any value of x in the displayed range. For example, Fig. 1.5 shows the graphs of the sine and cosine functions, that is, respectively:

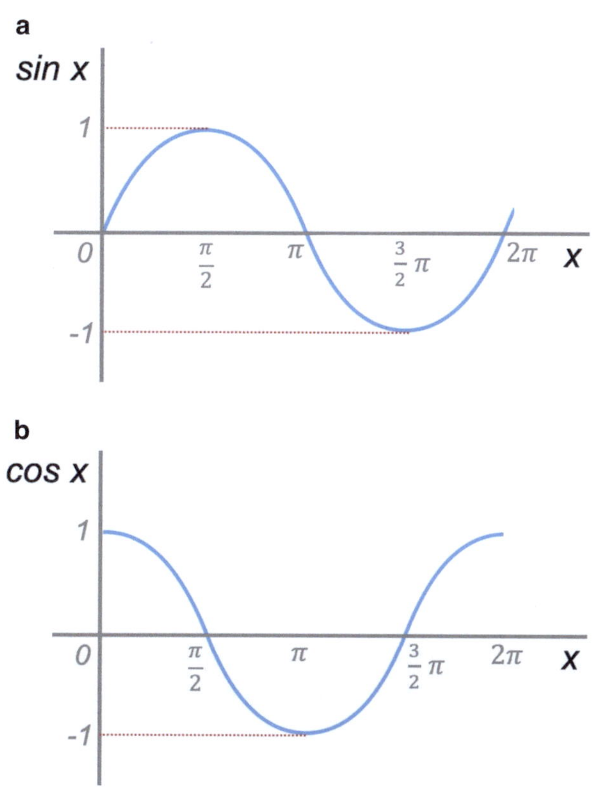

Fig. 1.5 Graphs of the sine function (**a**) and of the cosine function (**b**). The angles are expressed in radians ($\pi/2 = 90°$; π=180°; $3\pi/2 = 270°$; $2\pi = 360°$). As shown, sin x and cos x only assume values between −1 and 1. The sine of an angle increases with the angle width for angles between 0 and 90° and between 270 and 360°, while it decreases as the angle width increases for angles between 90 and 270°. Conversely, the cosine of an angle decreases as the angle width increases for angles between 0 and 180°, and increases with the angle width for angles between 180 and 360°

$$y = \sin x \tag{1.43}$$

and

$$y = \cos x \tag{1.44}$$

These functions (and the related graphs) represent the mathematical description of a sinusoidal wave: accordingly, we will encounter them several times throughout the book (e.g., in Chap. 16, when discussing electromagnetic waves, light, and pulse oximetry; in Chap. 19, when talking about mechanical waves, natural frequency, and invasive pressure monitoring; and in Chap. 20, which addresses another topic which has to do with waves, i.e., the Doppler effect).

1.4.1 Limits and Derivatives: The "True" Definition of Velocity, Gradients, and Other Quantities

Let us now briefly recall the notion of derivative of a function with a practical example (here too we will not go into much detail of mathematical formalism). Imagine driving down a straight road for a few minutes, and that your position x along the road at all times is described by Fig. 1.6. In other words, we can say that Fig. 1.6 shows the graph of the function:

$$x = f(t) \tag{1.45}$$

that is, the graph of the position x as a function of time t.

If you cover a total distance Δx_{tot} over the time Δt_{tot}, your *average velocity* is:

$$v_{avg} = \frac{\Delta x_{tot}}{\Delta t_{tot}} \tag{1.46}$$

Looking at Fig. 1.6a, it can easily be realized that Δx_{tot} and Δt_{tot} represent two sides of a right triangle whose hypotenuse is the segment OE: accordingly, remembering the definition of tangent of an angle (see Eq. 1.25 and Fig. 1.2), the ratio $\Delta x_{tot}/\Delta t_{tot}$ (i.e., the average velocity) is none other than the trigonometric tangent of the angle ϑ that the straight line joining the points O and E (which is called a *secant*) makes with the time (abscissa) axis.

In general, the average velocity in any tract AB of the graph (see Fig. 1.6a) is the ratio of the displacement Δx which occurs over a certain time interval Δt to Δt itself:

$$v_{avg} = \frac{\Delta x}{\Delta t} = \frac{x_2 - x_1}{t_2 - t_1} \tag{1.47}$$

Similarly to what we have said for the entire route, v_{avg} during the time interval Δt ($=t_2-t_1$) is equal to the tangent of the angle θ that the secant AB makes with the time axis (which is called the *angular coefficient* of the straight line AB). As can be seen from Fig. 1.6a (looking at it very carefully!), the angle θ is (a little bit) greater than the angle ϑ: accordingly, it is $\tan\theta > \tan\vartheta$, that is, the average velocity during the time interval Δt is higher than the average velocity over the entire route.

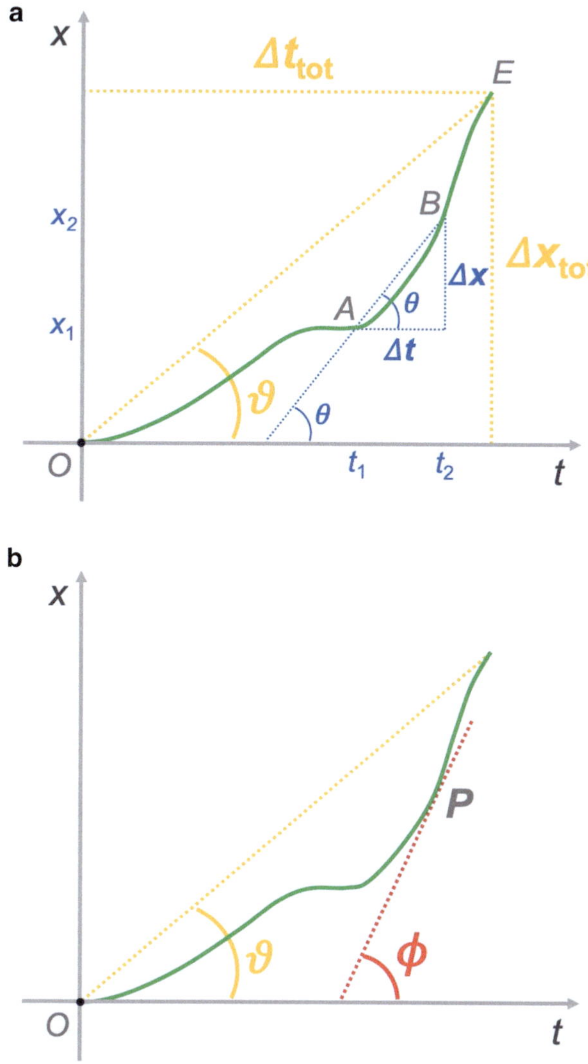

Fig. 1.6 A graph of the position x of an object moving with variable velocity in a straight line (e.g., a car along a road) as a function of time t. (**a**) Overall average velocity is the angular coefficient of the secant OE, i.e., $\Delta x_{tot}/\Delta t_{tot} = \tan \vartheta$; similarly, average velocity in the tract AB is the angular coefficient of the secant AB, i.e., $\Delta x/\Delta t = \tan \theta$. (**b**) Instantaneous velocity at the point P is the angular coefficient of the tangent at the point P, i.e., $\tan \phi$

We can calculate the velocity v at a single instant t (*instantaneous velocity*) as the limiting value to which the ratio $\Delta x/\Delta t$ approaches for values of Δt which become smaller and smaller, getting close to 0. In mathematics, this means finding the *limit* of the ratio $\Delta x/\Delta t$ for Δt tending to 0:

$$v = \lim_{\Delta t \to 0} \frac{\Delta x}{\Delta t} = \frac{dx}{dt} \tag{1.48}$$

where the notation dx/dt indicates the *derivative* of x with respect to t. As shown in Fig. 1.6b, the "geometrical meaning" of the derivative (and, in this case, of instantaneous velocity) is quite similar to what we have seen for average velocity, with the difference that we are now considering a single instant rather than a time interval (and, accordingly, a single point *P* along the graph rather than a tract *AB*); hence, we have to replace the secant *AB* with the *tangent* at point *P* (in practice, when Δt tends to 0, the tract *AB* tends to become a single point *P*, and the secant AB tends to coincide with the tangent at the point *P*; that is, as they say mathematicians [9], "the tangent is the limit of the secant"). The derivative of a function *at a point P* is, therefore, the (trigonometric) tangent of the angle ϕ that the tangent line at the point *P* makes with the abscissa axis or, in other words, the angular coefficient (i.e., the *slope*) of the tangent at the point *P*.

Generally speaking, the derivative of a function tells us "how fast" the function grows (or decreases) as its independent variable varies, and this is why the derivative of the curve describing the position of an object as a function of time represents the object's velocity. Usually, the term "velocity" always refers to instantaneous velocity (i.e., to the derivative of displacement with respect to time). Only when velocity is constant (i.e., it does not change over time), we can simply regard it as the displacement an object undergoes divided by the time required for such a displacement to occur; in this case, in fact, instantaneous velocity and average velocity are the same thing. As we will see throughout the book, many other physical quantities which are commonly roughly expressed by an "incremental ratio" are more generally and strictly defined in terms of limits and derivatives. For example, we usually say that the *density* of an object is the ratio of its mass to its volume, or the mass per unit volume; however, this is true only if density is *uniform*; otherwise, it is more correct to say that density is the limit of the ratio between mass and volume for an infinitesimal volume, i.e., for a volume which "tends to zero" (see Chap. 9). Similarly, you will have no surprise in reading, in Chap. 4, that "the derivative of concentration with respect to distance" is a more rigorous way to define a *concentration gradient* as compared with "the difference in concentration between two points divided by the distance between them," or, in Chap. 11, that electric current is better defined as "the derivative of electric charge by time" rather than "the charge which crosses any section of a conductor per unit time" (this because is not for sure that concentration varies evenly over space, or that the number of charges which pass through the conductor is constant over time).

This is all you will need to remember about differential calculus.

1.4.2 Integral Calculus: Distances, Areas, and Cardiac Output

The other branch of calculus, which is called integral calculus, includes two different types of "operations": *indefinite integrals* and *definite integrals*.

Roughly, the indefinite integral can be regarded to as the inverse of the derivative. So far, we have talked about the derivative of a function as something which is calculated *at a certain point*. However, the term "derivative" also refers to *a function* which maps the derivative of a function at every point. That is, given a function $f(x)$, it is often possible to define a function $f'(\mathrm{x})$ which gives the value of the derivative of $f(x)$ for any value of x. We call the function $f'(x)$ the derivative of $f(x)$. Without going into too many details, the indefinite integral of $f(x)$ is a function whose derivative is $f(x)$ itself. In other words, if $f'(x)$ is the derivative of $f(x)$, then the indefinite integral of $f'(x)$ is $f(x)$.

Perhaps it is better to give a practical example (which will lead us to the reason why we are mentioning this topic here). As discussed above, Fig. 1.6 shows the graph of the position x of your car along a road as a function of time t, i.e., the graph of the function $x = f(t)$. At the instant t_1, the car is located at the position x_1; at the instant t_2 it is at the position x_2; and so on (Fig. 1.6a). Your instantaneous velocity at the instant t_2 is, based on what we said, the angular coefficient of the tangent at point B (not shown) identified by the coordinates t_2, x_2, that is the derivative of the function $x = f(t)$ at $t = t_2$. We can build another graph, which displays the values of the derivative of the function $x = f(t)$ at every instant t, i.e., the values of velocity at every instant: this will be the graph of the derivative of the function $x = f(t)$, that is the graph of the function of velocity with respect to time. As we will see in Chap. 20, thanks to ultrasounds and Doppler effect, we can measure systolic blood velocity and build a curve of such a velocity over time (which can be visualized on the ultrasound machine screen). And here is the important point: since velocity is the derivative of the function of position with respect to time, the (indefinite) integral of the function of velocity with respect to time (which is referred to as the "velocity-time integral," or VTI, in echocardiography textbooks) is the function of position with respect to time. Accordingly, the VTI is none other than the set of all positions of a single red blood cell over the duration of a systole (i.e., the *distance* covered by it during this time period) and, once multiplied by the cross-sectional area of the left ventricle outflow tract (according to Eq. 1.20), it provides the volume of blood ejected into the aorta at each systole, that is the stroke volume. This is how cardiac output (CO) is measured by cardiac Doppler ultrasound or, similarly, by some minimally invasive or noninvasive CO monitors (see Chap. 20).

Finally, the definite integral allows to calculate the *area under a curve* (Fig. 1.7) [9]. Imagine you want to calculate the area A of the part of the Cartesian plane which is bounded below by the segment of the x-axis comprised between the coordinates c and d (whose length is d-c), above by the graph of the function $y = f(x)$, and on the sides by the perpendiculars to the x-axis at the points c and d (Fig. 1.7a). We can proceed in the following way.

First, we divide the interval d-c in a number n of equal subintervals, whose width is, evidently:

$$\Delta x = \frac{d-c}{n} \tag{1.49}$$

As shown in Fig. 1.7b, we can consider the area A, as a first approximation, as the sum S of the areas of the n rectangles whose base is Δx and whose height is the value of $f(x)$ for x equal to the right end of each subinterval:

$$S = f(x_1)\Delta x + f(x_2)\Delta x + \ldots + f(x_n)\Delta x \tag{1.50}$$

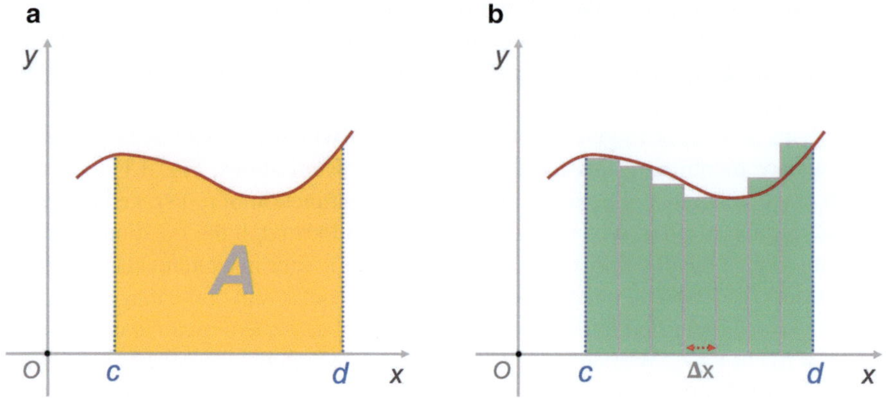

Fig. 1.7 Definite integral from c to d of the function y = *f*(*x*). **a**, **b**: see text

where $x_1, x_2, \ldots, x_n$ are the values of the right end of the first, second, …, *n*-th subinterval, respectively. As we will see in Chapters 4 and 10, the sum of a sequence of *n* elements (which is called a summation) can be abbreviated as follows:

$$S = \sum_{i=1}^{n} f(x_i)\,\Delta x \tag{1.51}$$

which is read "the summation for *i* ranging from 1 to *n* of the products $f(x_i)\ \Delta x$" (which are, evidently, the areas of the *n* rectangles).

As you can easily guess, the summation *S* progressively approaches the "true" area *A* as *n* becomes higher and higher (and, accordingly, Δx becomes narrower and narrower). Similarly to what we did for the derivative, we can then say that the area *A* is the limit of the summation *S* for Δx tending to 0 (or for *n* tending to infinite), which is called the definite integral of the function *f*(*x*) from *c* to *d*:

$$A = \lim_{n \to \infty} \sum_{i=1}^{n} f(x_i)\,\Delta x = \int_{c}^{d} f(x)\,dx \tag{1.52}$$

We will run into the need for calculating the area under a certain curve and, accordingly, into the symbol $\int$ of the integral a couple of times throughout the book. In particular, when we derive the modified *Stewart-Hamilton equation* for the measurement of cardiac output according to the thermodilution method (Chap. 10).

I think this is all the mathematics we will need to confidently tackle the book (you can find some other little mathematical suggestion directly in the chapters when needed). So, it is time for physics!

References

1. Barrow JD. The constants of nature. New York: Vintage; 2003.
2. Wigner E. The unreasonable effectiveness of mathematics in the natural sciences. Commun Pure Appl Math. 1960;13:1–14.

3. Tegmark M. Our mathematical universe: my quest for the ultimate nature of reality. London: Penguin; 2015.
4. Italian Committee for the Investigation of Claims of Pseudosciences (CICAP) website. https://www.cicap.org/n/articolo.php?id=200119. Accessed 18 Jan 2021.
5. Du Sautoy M. How to count to infinity. London: Quercus; 2020.
6. Susskind L, Hrabovsky G. The theoretical minimum: what you need to know to start doing physics. New York: Basic Books; 2014.
7. Piton G, Capellier G, Winiszewski H. Ultrasound-guided vessel puncture: calling for Pythagoras' help. Crit Care. 2018;22(1):292.
8. De Cassai A, Sandei L, Carron M. Trigonometry in daily ultrasound practice. Crit Care. 2018;22(1):356.
9. Courant R, Robbins H. What is mathematics? An elementary approach to ideas and methods, 2nd ed. revised by Ian Stewart. Oxford: Oxford University Press; 1996.

Part II

Gases, Bubbles and Surroundings

Coffee, Popcorn, and Oxygen Cylinders: The Ideal Gas Law

2

Contents

Where's Physics

Daily life	How does a "moka" coffee pot work Popcorn Earache during flight Troublesome passengers on the aircraft
Physics involved	Ideal gas law (and its special cases: Boyle's law and Gay-Lussac's laws)
Clinical practice	Calculating the duration of an oxygen cylinder Decompression illness/hyperbaric therapy Tracheal tube cuff and hypothermia Tracheal tube cuff and altitude (helicopter transportation of intubated patients)

A. Pisano, *Physics for Anesthesiologists and Intensivists*,
https://doi.org/10.1007/978-3-030-72047-6_2

2.1 Strange Associations

Figure 2.1 shows some coffee coming out of a "moka" coffee pot (a); fragrant popcorn (b); an oxygen cylinder (c); a tracheal tube with its "cuff" in plain sight (d); a diver (e); and an airplane that is gaining altitude (f). Seemingly, these pictures have little or nothing in common with each other (although I found recipes of coffee with popcorn on the internet). However, all of them are connected by at least one thing: physics! In particular, the so-called *ideal gas law* correlates pressure, volume, and temperature of a gas with one another. As discussed below, the moka coffee maker

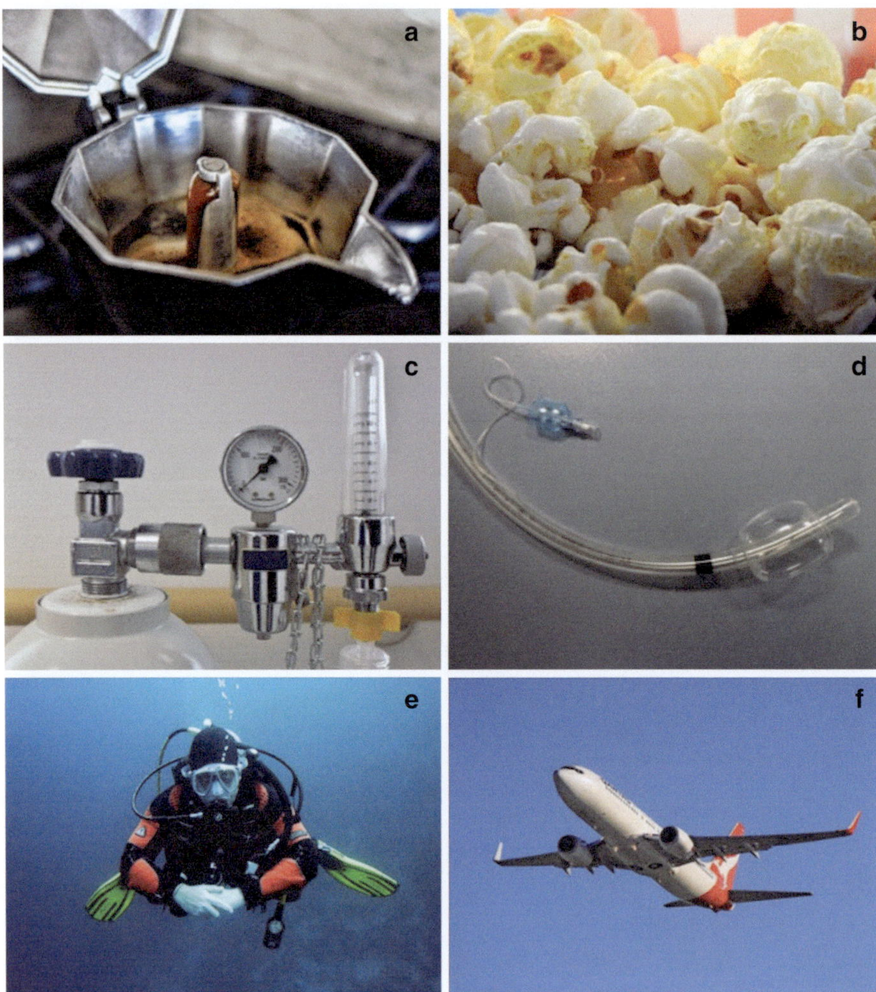

Fig. 2.1 Some objects and situations that apparently have little or nothing in common with each other; actually, all of them have something to do with gas laws (see text)

works according to this law and popcorn is also made thanks to it; the same physical principle allows to calculate the reserve of oxygen in a cylinder based on the pressure inside it; similarly, volume and pressure of a tracheal tube cuff are affected by both temperature and altitude and, accordingly, adjustments of cuff inflation may be needed (at least theoretically) during intraoperative deep hypothermia or helicopter transportation of intubated patients in order to avoid leakage or tracheal injury, respectively; finally, divers must be very familiar with the behavior of gases to avoid decompression illness, and pressure and volume changes similar to those causing this potentially severe disease during a dive may also be responsible for earache (or other more embarrassing troubles) during an airplane flight.

2.2 Delicious Scents and a Nauseating Stench

The Italian "espresso" coffee pot, also called *moka* (Fig. 2.2), is known around the world. In Italy, there is at least one in every kitchen. However, very few people really know how it works. All those to whom I asked answered that water flows through the funnel, which contains the ground coffee, after *boiling*. Indeed, this seems a widely held belief [1]. Figure 2.2b shows what happens in reality: *as temperature rises*, the *pressure* of the mixture of air and water vapor trapped between the water surface and the funnel increases (soon we will see why); when this pressure overcomes the sum of the atmospheric pressure (at sea level, 760 mmHg or 101.3 kPa) and the pressure of filtration (which depends on physical properties of

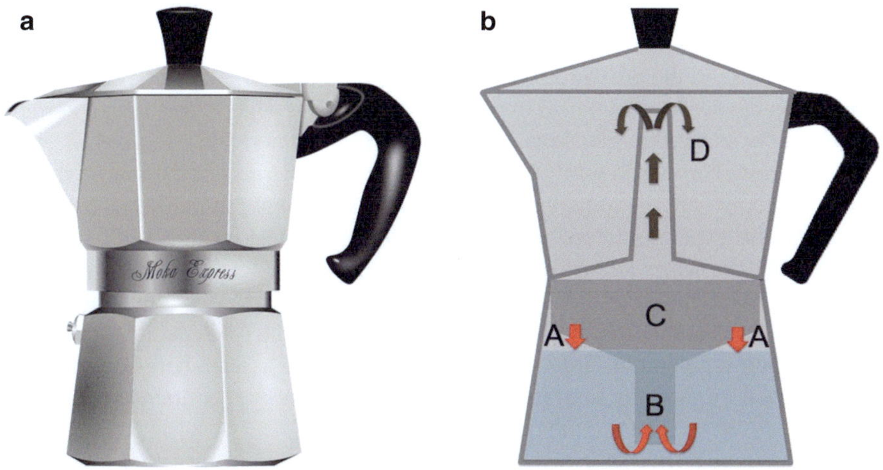

Fig. 2.2 A *moka* coffee pot (**a**) and a schematic representation of how it works (**b**). According to the Gay-Lussac's law (see Sect. 2.3.3), as temperature rises the pressure of the mixture of air and water vapor trapped between the water surface and the funnel increases (*A*), pushing water into the funnel (*B*) and through the ground coffee which is contained in the top of it (*C*). The coffee then comes out from the nozzle in the middle of the pot (*D*) after aromatic compounds have been dissolved in hot water

both water and ground coffee and on the funnel dimensions, according to the *Darcy's law of linear filtration*), the hot water is pushed into the funnel and through the ground coffee, from which it extracts aromatic compounds [2]; the coffee then comes out from the nozzle in the middle of the pot. Only the small amount of water remaining at the end will boil (if the pot is not readily removed from the heat) due to the sudden decrease in pressure after the coffee came out (see Box 2.1 and Chap. 15).

Box 2.1: Atmospheric Pressure and Boiling (See Also Chap. 15)

A liquid boils when its *saturated vapor pressure* (see Sect. 15.1.1), which depends only on temperature (for a given substance), equals the external atmospheric pressure. Saturated vapor pressure (SVP) of water at 20 °C (or 293 K) is 17.5 mmHg (or 2.3 kPa). When you put water on the stove to cook pasta, its vapor pressure increases with temperature. At 100 °C (373 K), water vapor pressure reaches 760 mmHg (or 101.3 kPa), that is the atmospheric pressure at sea level. Accordingly, water boils at 100 °C at sea level. In the mountains, where atmospheric pressure is lower, water needs to reach a lower SVP (and, accordingly, a lower temperature) to boil. This explains the statement "water will boil due to the sudden decrease of pressure" (in fact, as long as the coffee does not come out, pressure above the water in the pot is higher than atmospheric pressure due to the accumulating steam). Moreover, it explains why pasta is cooked badly in the high mountains (cooking temperature is lower)!

The increase in the pressure of a gas with temperature (when the "space" which it occupies cannot increase) also explains the explosion which turns maize kernels into the tasty snack that gladdens our evenings at the cinema [3]. Unlike all other cereals (whose kernels do not "pop" when heated), the kernels of the maize variety called "popcorn" are the only ones having a hard and nonporous shell. As temperature rises, the water contained inside the grain evaporates, turning itself in water vapor. Since the water vapor is trapped within the grain due to the very resistant shell, its pressure increases with the further increase in temperature, up to nearly 10 atmospheres (10 atm = 7600 mmHg ≅ 1013 kPa, that is the pressure at 90 m underwater!) when temperature approaches 180 °C (~360 °F). The shell is eventually overwhelmed by the very high internal pressure and starches contained inside the grain are expelled violently: popcorn is ready!

Now let's move from the sweet aromas of coffee and popcorn to a much less pleasant smell. It can be very annoying to be seated on an airplane next to a passenger who suffers from flatulence. This is not only because the space on the airplane is limited, but also because his/her disorder might be accentuated during the flight due to the increase of intestinal gas volume caused by the reduction of atmospheric pressure with altitude (see Chap. 9). Indeed, among the many gastrointestinal problems which can occur in the high mountains [4] and, although to a lesser extent (due to cabin pressurization), on an airplane, high altitude flatus expulsion (HAFE) syndrome seems to be related to the effect of pressure on gas volume [5].

As mentioned, these are all examples of how pressure, volume, and temperature of gases influence each other according to the *ideal gas law* and its special cases, which are discussed below.

2.3 Ideal Gas Law

Pressure, volume, and temperature of an *ideal* gas are related by the following equation:

$$pV = nRT \tag{2.1}$$

where p is absolute pressure, V is volume, n is the number of moles of gas, R is the so-called *universal gas constant* (which has the same value for all gases, i.e., 8.31 J/mol K), and T is the temperature in kelvins (K).

One mole (1 mol) of a given substance contains 6.02×10^{23} atoms or molecules of that substance. This number is called the *Avogadro's number* (N_A). Accordingly, the number of molecules N present in a sample of a given substance is the product of N_A and the number of moles n:

$$N = n N_A \tag{2.2}$$

Equation (2.1) (*ideal gas law* or *universal gas law*) can therefore be rewritten for the number of molecules N (instead of for the number of moles n) as follows:

$$pV = \frac{N}{N_A} RT \tag{2.3}$$

$$pV = NkT \tag{2.4}$$

where $k = R/N_A$ is the so-called Boltzmann constant (1.38×10^{-23} J/K).

All *real* gases such as oxygen, as well as any mixture of different gases (not interacting each other) such as air follow the ideal gas law, provided that their *density* (see Chap. 9) is sufficiently low, so that the volume of each molecule is negligible compared to the volume which is occupied by the gas.

All other gas laws can be easily derived from Eqs. (2.1) or (2.4) as special cases.

2.3.1 Boyle's Law

If temperature remains constant, volume and pressure of a sample of any gas are inversely proportional.

In fact, since R (or k) is a constant and the quantity of gas (i.e., the number of moles n or the number of molecules N) is the same, if also temperature does not change the product nRT (or NkT) is a constant and (2.1) or (2.4) can be written as follows:

$$pV = a\,\text{constant} \tag{2.5}$$

This is what happens to intestinal gases when altitude increases: as pressure decreases, volume must increase (and flatulence may worsen).

Boyle's law also explains *earache* which can occur during an airplane flight (especially during descent). As altitude rises in the ascent phase, atmospheric pressure decreases (according to the so-called *law of atmospheres*, which is discussed in Chap. 9). Cabin pressure decreases too (despite pressurization) and, according to Boyle's law, the volume occupied by the air contained in the middle ear tends to increase. As a consequence, part of this air escapes through the Eustachian tube, thus equalizing pressure across the tympanic membrane (since the volume V of the middle ear does not change, and assuming that also temperature T remains constant, a reduced pressure in the middle ear must imply a lower number of molecules N of air within it in order to satisfy Eq. 2.4). During descent, cabin pressure and, accordingly, the pressure in the ear canal, progressively increases; in order to equalize again the pressure across the tympanic membrane, some molecules of air should now return into the middle ear through the Eustachian tube. However, this can be more difficult as compared with the reverse path, especially if the Eustachian tube is congested, for example, due to upper airway infections such as common cold (the reason for this involves other physical laws and will become clearer in the last section of this chapter when discussing about what may happen to the tracheal tube cuff during helicopter transportation of intubated patients). The pressure difference, not promptly zeroed, across the tympanic membrane causes pain, which can be relieved by maneuvers that help Eustachian tube opening, such as swallowing or chewing a gum.

2.3.2 First Law of Gay-Lussac (or Charles's Law)

If pressure remains constant, the volume of a given amount of gas (for example, contained in a deformable container such as a balloon) is directly proportional to its temperature:

$$V / T = a\,\text{constant or}\,V \propto T \tag{2.6}$$

This law, together with *Archimedes' principle*, explains why a hot air balloon flies (see Chap. 3).

2.3.3 Second Law of Gay-Lussac (or, Simply, Gay-Lussac's Law)

If volume remains constant, the pressure of a given amount of gas is directly proportional to its temperature:

$$p / T = a\ \text{constant or}\ p \propto T \tag{2.7}$$

This is the case of the coffee pot (see Fig. 2.2) and of the popcorn grain. A gas, particularly a mixture of air and water vapor in the coffee pot and water vapor in the popcorn grain, is trapped in a closed and non-deformable space (i.e., the volume is constant); according to Gay-Lussac's law, its pressure increases as temperature rises.

From a thermodynamic standpoint, the pressure of a gas in a container reflects the number of collisions of the gas molecules with the container walls (*kinetic theory of gases*). In particular, pressure is proportional to the so-called root-mean-square speed (v_{RMS}) of the molecules, which in turn depends on temperature, according to the following relationship:

$$v_{RMS} = \sqrt{\frac{3RT}{M}} \tag{2.8}$$

where T is temperature, M is molar mass (i.e., the mass of one mole), and R the universal gas constant. So, higher temperature means faster molecules and, accordingly, more collisions, i.e., higher pressure. It can be said, indeed, that temperature of a given substance is a "measure" of the *kinetic energy* (E_K) of its molecules (see Box 2.2).

Box 2.2: Temperature and Kinetic Energy: A Look at the Kinetic Theory of Gases
According to the "kinetic theory of gases," volume, pressure, and temperature of a gas are all related to the motion of its molecules. A gas can be considered as a whole of tiny balls (whose volume is negligible with respect to the volume occupied by the gas) which continuously move, with different speeds, bumping with each other and with the container walls. The "average speed" of the molecules is expressed as the square root of the mean of the squares of each speed, i.e., the so-called root-mean-square speed (v_{RMS}).

In general, any object (whose mass is m) which moves with speed v has a *kinetic energy* (E_K) that is equal to:

$$E_K = \frac{1}{2}mv^2$$

Similarly, the average kinetic energy of a gas molecule with molecular mass m is:

$$\overline{E}_K = \frac{1}{2}m\left(v_{RMS}\right)^2$$

After substituting for v_{RMS} from Eq. (2.8), the above equation becomes:

$$\overline{E}_K = \frac{1}{2}m\frac{3RT}{M}$$

where T is temperature, M is molar mass, and R is the universal gas constant. Remembering that M/m is the Avogadro's number N_A (i.e., the mass of a mole divided by the mass of a molecule is equal to the number of molecules which are contained in a mole), we can write:

$$\overline{E}_K = \frac{1}{2}\frac{3RT}{N_A}$$

and:

$$\overline{E}_K = \frac{3}{2}KT$$

where $k = R/N_A$ is the Boltzmann constant.

It clearly results from the last equation that the temperature of a gas is nothing different from the average kinetic energy of its molecules.

2.3.4 Avogadro's Law

Two equal volumes of any gas at the same pressure and temperature hold the same number of molecules. Just look Eq. (2.4) (if p, V, and T do not change and k is a constant then N can assume only one value).

2.3.5 Dalton's Law

The total pressure exerted by a mixture of non-reactive gases is equal to the sum of the partial pressures of the individual gases. This law is discussed in details in Chap. 4.

2.4 Calculating the Duration of an Oxygen Cylinder

Some anesthesiologists may have happened to transfer a patient on mechanical ventilation (or oxygen therapy) and to be worried that the oxygen cylinder could be emptied too early. As an example, a patient breathing through a Venturi mask (see Chap. 7) delivering an inspiratory oxygen fraction (FiO_2) of 0.6 (or 60%) needs an oxygen flow of 15 liters per minute (L/min). If the transfer takes 1 h, will a full cylinder with 10 L capacity be enough? And the same cylinder half full? And what about a 5 L cylinder?

Common cylinders for hospital or home use contain oxygen in the gas phase. In fact, oxygen can not be liquefied at room temperature because its *critical temperature* (namely, the temperature above which a substance cannot exist in the liquid phase, regardless of the pressure applied) is very low (−118.57 °C or 154.58 K) [6].

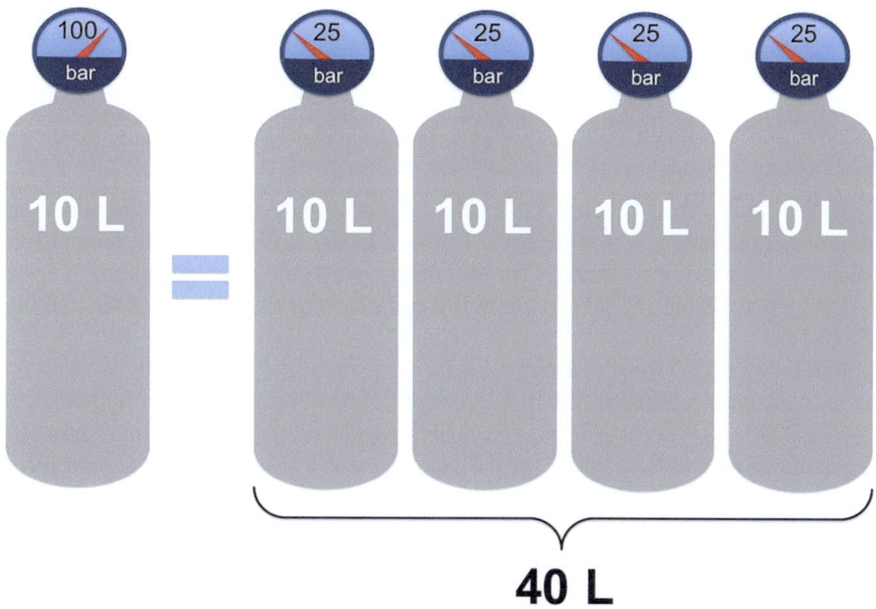

Fig. 2.3 Boyle's law. At a fixed temperature, the same amount of any gas which is contained in a 10 L cylinder at a pressure of 100 bar will occupy four cylinders identical to the first (or 40 L) at a pressure of 25 bar, so that the product of pressure and volume remains constant ($10 \times 100 = 25 \times 40$)

As we will see, this is a fortune: in fact, it is not possible to know the content of a cylinder which contains a liquefied gas without a weighing scale!

For 1 h, the abovementioned patient needs:

$$15 \text{ L min}^{-1} \times 60 \text{ min} = 900 \text{ L} \tag{2.9}$$

of oxygen. How many liters of oxygen are contained in a full 10 L cylinder? The answer is very simple: 10 L. And if it is half full, or almost empty? Again 10 L! Unlike solids and liquids, which have a *definite volume* (the former, but not the latter, also have a defined shape), gases always occupy all the volume that they have available. Accordingly, it is not enough to know the volume of a gas to know its "quantity."

It is clear from Eqs. (2.1) or (2.4) that, at a given temperature, the number of moles (or molecules, respectively) which are contained in a certain volume of gas depends on its pressure. Indeed, the "quantity" of oxygen which is contained in a cylinder is usually indicated by a pressure gauge and expressed in *bar* (1 bar ≈ 1 atm = 760 mmHg = 101.3 kPa). Oxygen cylinders are commonly filled up to 200 bar.

According to Boyle's law (Eq. 2.5), if a given amount of gas occupies the volume V_1 at a pressure p_1, it will occupy the volume V_2 (> V_1) at a pressure $p_1 < p_2$ (provided that temperature does not change), so that the product of its volume and pressure remains constant (see Fig. 2.3):

$$p_1 V_1 = p_2 V_2 \tag{2.10}$$

The volume V_2 can be easily calculated as follows:

$$V_2 = \frac{p_1 V_1}{p_2} \tag{2.11}$$

In the case of the oxygen cylinder, the volume V_2 which can be delivered to the patient *at atmospheric pressure* ($p_2 = 1$ atm ≈ 1 bar) is the product of the cylinder capacity (V_1) and the pressure p_1 (in bar) displayed on the pressure gauge:

Oxygen available (*L*) = Pressure in the cylinder (*bar*) x Capacity of the cylinder (*L*) (2.12)

that is Eq. (2.11) for $p_2 = 1$ bar.

In the example above, a full 10 L cylinder (which has a pressure of 200 bar) will release 200 × 10 = 2000 L of oxygen, which will be certainly sufficient to meet the needs of our patient (i.e., 900 L for 1 h). Conversely, a half-full 10 L cylinder (pressure = 100 bar) will release 100 × 10 = 1000 L, as well as a full 5 L cylinder (200 × 5 = 1000); in both cases, I would be quite concerned! First, because the patient would have an oxygen reserve for only a few minutes more than expected (100 L/15 L $min^{-1} \approx 6$ min), and most importantly, because it is advisable to never fully rely on the accuracy of these pressure gauges and, accordingly, to approximate by (large) defect the amount of available oxygen.

However, oxygen cylinders with a digital display which indicates the remaining "liters" by incorporating the calculation of the product of pressure and capacity are now available in order to help physicians ... to forget physics!

In the case of cylinders containing a liquefied gas, e.g., carbon dioxide (CO_2), the pressure measured by the gauge does not correlate with the amount of the substance inside the cylinder. In fact, the liquid CO_2 is in equilibrium with its gaseous phase, whose pressure in these circumstances is always the *saturated vapor pressure* of CO_2 at the current temperature (see Chap. 15), whatever the amount of liquid contained in the cylinder. Accordingly, as mentioned, you should weigh the cylinder (and know its weight when full) in order to estimate the remaining amount of CO_2. The same applies to nitrous oxide.

2.5 Decompression Illness and Hyperbaric Therapy

Together with *Henry's law* (which is discussed in Chap. 5), Boyle's law has a key role in both the pathophysiology of decompression illness (DCI) and its recompression therapy with hyperbaric oxygen (HBO). A detailed discussion of these topics is beyond the scope of the present book and can be found in Vann et al. [7] In a few words, DCI includes two different diseases, both as a consequence of a reduction in environmental pressure (such as during the ascent of a diver): arterial gas embolism (AGE) and decompression sickness (DCS). While AGE occurs as a direct consequence of Boyle's law, since the reduced ambient pressure leads to expansion of alveolar gas causing pulmonary barotrauma and allowing gas to enter the arterial circulation, DCS involves the formation of bubbles in both tissues and blood due to Henry's law (see Chap. 5), followed by the increase in the size of such bubbles with decompression, according to Boyle's law. HBO therapy is the standard treatment for DCI [8], in which it acts primarily by reducing bubbles volume, thanks to the increase in the environmental pressure (according, again, to Boyle's law). Moreover, a couple of practical issues in the care of patients with DCI or who need HBO therapy result from Boyle's law. The first is that a patient with DCI who needs urgent recompression can be transferred to a hyperbaric chamber by helicopter, provided to fly at low altitude (in order to avoid a further increase in the size of bubbles which would occur at higher altitudes due to a significant reduction in atmospheric pressure) [7]. Finally, in an intubated patient requiring HBO therapy, the endotracheal tube cuff balloon should be filled with saline (which is substantially incompressible) instead of air in order to avoid the reduction of its volume due to the increasing pressure [9].

2.6 Gas Laws and the Tracheal Tube Cuff

Hyperbaric therapy is not the only clinical setting in which the inflation of the tracheal tube cuff may be affected by either environmental or physiological changes according to the above discussed gas laws, leading to the possible need for adjustments in order to avoid leakage or tracheal injury.

As anesthesiologists and intensivists are well aware, the inflation pressure of the cuff of an endotracheal (or tracheotomy) tube should be between 20 and 30 cm H_2O (or 15–22 mmHg) [10, 11]. Pressures below 20 cmH_2O may allow microaspiration and, accordingly, increase the risk of ventilator-associated pneumonia (VAP). Conversely, pressures greater than 30 cmH_2O may compromise tracheal wall capillary blood flow and are associated to an increased risk of tracheal complications such as ulcerations, tracheomalacia, stenosis, and even rupture or tracheo-esophageal fistulae.

As an air-filled system, the tracheal tube cuff is subject to the ideal gas law. For example, as *temperature* (of the air contained in the cuff) drops, both cuff volume and intra-cuff pressure tend to decrease according to Eq. (2.1). Even assuming that the reduction of cuff volume with temperature is such as to leave pressure unchanged (that is, the situation described by Charles's law occurs), this will anyhow lead to a drop in intra-cuff pressure. In fact, the intra-cuff pressure that we measure after injecting a certain amount of air in the cuff is the result of the interaction between the cuff and the tracheal wall, which opposes to the increase in cuff volume by pushing on the cuff as the cuff pushes on tracheal wall itself while trying to expand, according to the *Newton's third law* (see Chap. 14, Sect. 14.1.3). In order to realize this, just measure cuff pressure after injecting the "usual" amount of air in the cuff of a tube which has not been inserted in trachea but is held in the hands: as the cuff is free to expand, measured pressure will be very low. Theoretically, *hypothermia* which develops (or is induced) after the patient was intubated may reduce intra-cuff pressure so much to impair cuff sealing (evidently depending on the initial inflation pressure). A significantly reduced intra-cuff pressure has been reported in patients undergoing cardiac surgery with hypothermic cardiopulmonary bypass (CPB) or deep hypothermic circulatory arrest (DHCA) [12–14]. In particular, a small randomized controlled trial (RCT) of 24 patients undergoing pulmonary endarterectomy with DHCA (at 16–18 °C, or 61–64 °F) found that not checking and adjusting intra-cuff pressure during hypothermia was associated with a significant increase in the rate of silent aspiration (as assessed by transfer of methylene blue dye across the cuff) [14]. However, whether this implies a significantly increased risk of pulmonary complications is unclear [15].

The opposite situation occurs when transporting an intubated patient by helicopter (which, unlike an airliner, is not pressurized). As altitude increases, atmospheric pressure decreases exponentially (according, as mentioned, to the law of atmospheres, which is discussed in Chap. 9). Accordingly, the air contained in the cuff tends to expand (Boyle's law) and, as a consequence, the cuff will push harder on the tracheal wall, with possible increased risk of tracheal injury. In accordance with Newton's third law, the tracheal wall will push harder in the opposite direction, leading to an *increase* in intra-cuff pressure. Note that a similar mechanism explains why the air flow from the middle ear across the Eustachian tube during the airplane ascent is easier than the opposite pathway during descent (as discussed in Sect. 2.3.1): in fact, the significant increase in pressure which occurs during the ascent in the middle ear, since it is an enclosed space whose walls cannot expand, forces air through the only escape (i.e., the Eustachian tube); conversely, the pharynx is widely communicating with the outside and it is more difficult that air is channeled in a narrow conduit although the gradually increasing pressure gradient between the pharynx and the middle ear during descent will sooner or later impose the transfer of air molecules from the pharynx to the middle ear.

Although some observational investigations suggested a possible significant increase in intra-cuff pressure during helicopter emergency medical service (HEMS) transportation [16, 17], one study found no clinically relevant increases in intra-cuff pressure when flying below 3000 feet (about 900 m) [18]. However, in a series of 55 patients who were intubated before being referred to HEMS, Tennyson et al. [19] found that most patients (84%) had intra-cuff pressures (before flying) much higher than the maximum recommended. Evidently, helicopter flight would further increase the risk of tracheal injury in patients with already overinflated tracheal tube cuffs, and checking intra-cuff pressure with a pressure gauge (and adjusting it if needed) should be strongly recommended in patients undergoing HEMS transportation.

References

1. Navarini L, Nobile E, Pinto F, Scheri A, Suggi-Liverani F. Experimental investigation of steam pressure coffee extraction in a stove-top coffee maker. Appl Therm Eng. 2009;29:998–1004.
2. Gianino C. Experimental analysis of the Italian coffee pot "Moka". Am J Phys. 2007;75(1):43–7.
3. Czerski H. Storm in a teacup: the physics of everyday life. New York: W. W. Norton & Company; 2018. p. 13–5.

4. Anand AC, Sashindran VK, Mohan L. Gastrointestinal problems at high altitude. Trop Gastroenterol. 2006;27(4):147–53.
5. Auerbach P, Miller YE. High altitude flatus expulsion (HAFE). West J Med. 1981;134(2):173–4.
6. Wagner W, Ewers J, Pentermann W. New vapour-pressure measurements and a new rational vapour-pressure equation for oxygen. J Chem Therm. 1976;8(11):1049–60.
7. Vann RD, Butler FK, Mitchell SJ, Moon RE. Decompression illness. Lancet. 2011;377(9760):153–64.
8. Bennett MH, Lehm JP, Mitchell SJ, Wasiak J. Recompression and adjunctive therapy for decompression illness: a systematic review of randomized controlled trials. Anesth Analg. 2010;111(3):757–62.
9. Weaver LK. Hyperbaric oxygen in the critically ill. Crit Care Med. 2011;39(7):1784–91.
10. Pisano A. Pitfalls from physics: why we can't "feel" the tube cuff pressure with our fingers. Anesth Analg. 2017;124(4):1368.
11. Pisano A, Verniero L, Galdieri N, Corcione A. Assessing the correct inflation of the endotracheal tube cuff: a larger pilot balloon increases the sensitivity of the 'finger-pressure' technique, but it remains poorly reliable in clinical practice. J Clin Monit Comput. 2019;33(2):301–5.
12. Souza Neto EP, Piriou V, Durand PG, et al. Influence of temperature on tracheal tube cuff pressure during cardiac surgery. Acta Anaesthesiol Scand. 1999;43(3):333–7.
13. Kako H, Alkhatib O, Krishna SG, et al. Changes in intracuff pressure of a cuffed endotracheal tube during surgery for congenital heart disease using cardiopulmonary bypass. Paediatr Anaesth. 2015;25(7):705–10.
14. Rubes D, Klein AA, Lips M, et al. The effect of adjusting tracheal tube cuff pressure during deep hypothermic circulatory arrest: a randomised trial. Eur J Anaesthesiol. 2014;31(9):452–6.
15. De Hert S, De Baerdemaker L. Randomised controlled trials: are we looking at treatment effects or absence of good clinical practice in control groups? Eur J Anaesthesiol. 2014;31(9):450–1.
16. Delorenzo AJ, Shepherd M. Jennings PA. Endotracheal cuff pressure changes during helicopter transport: a systematic review. Air Med J. 2017;36(2):81–4.
17. Orsborn J, Graham J, Moss M, et al. Pediatric endotracheal tube cuff pressures during aeromedical transport. Pediatr Emerg Care. 2016;32(1):20–2.
18. Weisberg SN, McCall JC Jr, Tennyson J. Altitude-related change in endotracheal tube cuff pressures in helicopter EMS. West J Emerg Med. 2017;18(4):624–9.
19. Tennyson J, Ford-Webb T, Weisberg S, LeBlanc D. Endotracheal tube cuff pressures in patients intubated prior to helicopter EMS transport. West J Emerg Med. 2016;17(6):721–5.

Boats, Balloons, and Air Bubbles: Archimedes' Principle

3

Contents

Where's Physics

Daily life	Boats, ships, buoys, life jackets, and all other things that float on the sea (Flying) balloons for kids, ice cubes, hot air balloons Playing dead afloat, swimming in fresh or salt water
Physics involved	Archimedes' principle
Clinical practice	Protecting the brain from air embolism: • during cardiac surgery with cardiopulmonary bypass • when removing a central venous catheter Facilitate de-airing (open-heart surgery) Handle a syringe

3.1 Archimedes' Principle: Gravity Not Always Makes You Fall

Everyone knows that boats, ships, life jackets, buoys, and even the human body float in the sea. We also know that ice cubes float in our soft drink, that a balloon filled with helium flies upward (making children happy), as well as one filled with hot air (see Fig. 3.1). Moreover, many people are aware that swimming (or at least stay afloat) in the pool is a bit more exhausting than in the sea (provided that it is calm, of course). All of this is explained by the same physical law: *Archimedes' principle*.

A. Pisano, *Physics for Anesthesiologists and Intensivists*,
https://doi.org/10.1007/978-3-030-72047-6_3

It states that when a body is fully or partially submerged in a *fluid* (that is, in a liquid, in a gas, or in a gas mixture such as air), the fluid pushes it upward with a buoyant force that equals the weight of the fluid that has been displaced (Fig. 3.2).

Fig. 3.1 Some examples of the Archimedes' principle from daily life. All these objects have a lower density than the fluid (**a**, **d**: water; **b**, **c**: air) in which they are immersed (see text)

Fig. 3.2 The forces acting on a body of volume *V* and weight W that is submerged in a fluid. B, buoyant force (Archimedes' principle); m_f, mass of the volume *V* of fluid which is displaced by the body; d_f, fluid density; g, free-fall acceleration; m_b, mass of the body; d_b, density of the body

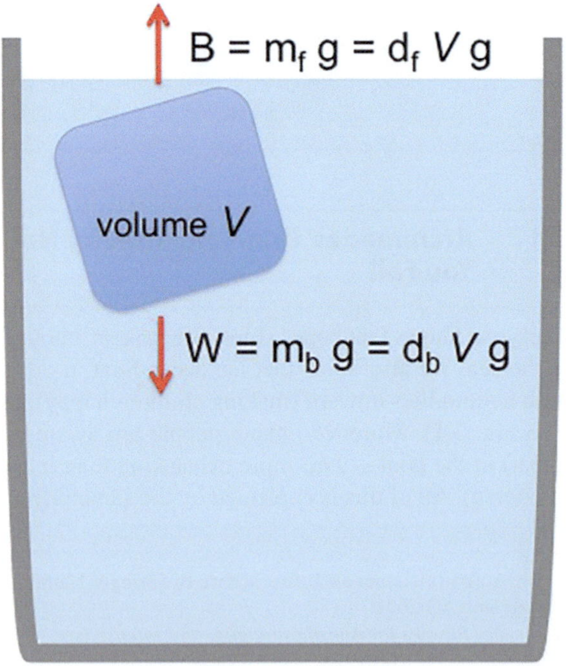

According to *Newton's second law*, the weight of a body is the product of its mass and free-fall acceleration (see Box 3.1 and Chap. 14). Therefore, the buoyant force B on a body in a fluid has magnitude:

$$B = m_{\mathrm{f}}\, g \tag{3.1}$$

where m_{f} is the mass of the fluid that has been pushed out of the way by the body and g is free-fall acceleration.

Remembering that density d of a body is, roughly, the mass per unit volume (see Chap. 9), i.e., the ratio between the mass m and the volume V of that body:

$$d = \frac{m}{V} \tag{3.2}$$

m can be written as the product of d and V ($m = dV$) and Eq. (3.1) becomes:

$$B = d_{\mathrm{f}}\, V\, g \tag{3.3}$$

where d_{f} is the fluid density and V is the volume of the fluid displaced, *which is clearly equivalent to the volume of the body itself (or of its submerged portion).*

While the buoyant force pushes the body upward, its weight pulls it down. According again to Newton's second law, the magnitude of the body weight W is:

$$W = m_{\mathrm{b}}\, g = d_{\mathrm{b}}\, V\, g \tag{3.4}$$

where m_{b} is the mass of the body and d_{b} is its density.

As you can see, the only difference in the calculation of the two forces (the buoyant force B and the weight W) is *density*: B is proportional to the density of the fluid, while W is proportional to the density of the body (see Fig. 3.2). The net force acting on the body is the difference between B and W (since the two forces have the same direction and opposite sign):

$$W - B = d_{\mathrm{b}}\, V\, g - d_{\mathrm{f}}\, V\, g = \left(d_{\mathrm{b}} - d_{\mathrm{f}}\right) V\, g \tag{3.5}$$

Accordingly, if the density of the body is lower than that of the fluid, the buoyant force will be greater than the body weight and the body will be pushed upward. Conversely, it will go down. If a body floats (i.e., it remains only partially submerged) it means that $B = W$.

Interestingly, the "fight" between the two forces W and B is not, as one might think, that between gravity and a different kind of force, but it is rather a "fight" of gravity against itself (due to the presence of the surrounding fluid). In fact, if we consider the example of an object submerged in a liquid, the buoyant force is due to the increase in *hydrostatic pressure* with depth (Stevin's law) (see Chap. 9). Since hydrostatic pressure is the pressure exerted by a column of liquid due to its weight, and weight is in turn an effect of gravity (see Box 3.1), it appears clear that, ultimately, also the buoyant force is due to gravity.

All the above-mentioned phenomena of everyday life are easily explained with differences in density. All objects filled with air (which is less dense than water) float in the sea: even the very heavy ships float thanks to their large air-filled spaces (Fig. 3.1a). Helium has a lower density as compared with air [1]: therefore, balloons filled with this gas go upward in the air (Fig. 3.1b). According to Charles's law (see Chap. 2), the volume of a gas increases as temperature rises: a hot air balloon is initially inflated by heating the air (Fig. 3.1c), which becomes less dense (due to the increase in the balloon volume); when the balloon reaches its maximum volume, some air escapes from the opening at the base as temperature continues to rise, thus leading to a further decrease in density (the same volume, less molecules). Unlike almost all other substances, water expands (and, accordingly, its density decreases) when it becomes solid: this is why ice cubes float in water (and even more in soft drinks which contain sugar and other substances) (Fig. 3.1d). Finally, salt water is more dense than fresh water: that's why you float more easily in the sea than in a swimming pool.

Box 3.1: Mass and Weight: Newton's Second law and the Principle of Equivalence (See Also Chap. 14)

Newton's second law states that the net force F acting on a body is the product of its mass m and its acceleration a (namely, the variation in speed over time):

$$F = m\,a$$

The mass of an object is generally described as the "quantity of matter" of which it is composed. However, Newton's second law provides a stricter definition of mass as the force that is needed to impart to a body an acceleration of unit value ($a = 1$). In other words, mass can be regarded as the "resistance" that a body opposes to a force which tries to make it accelerate. As discussed in Chap. 14, it follows from Newton's second law that, if a body accelerates (or, in other words, is subject to acceleration), this means that a net force is acting on it.

The weight W of a body (which is often confused with its mass m) is the *force* that is exerted on it by effect of gravity. According to Newton's second law, it is:

$$W = m\,g$$

where g is the free-fall acceleration (i.e., the acceleration due to gravity), that is about 9.8 m/s^2 (or 32 ft./s^2). On the moon, gravitational attraction is about 1/6 that on the Earth ($g \approx 1.6$ m/s$^2 \approx 5.2$ ft./s^2); accordingly, our weight would be six times less than on the Earth. However, our mass (i.e., the quantity of "matter" and, unfortunately, of fat) would be exactly the same!

For the accuracy, and for your curiosity, gravity is not a force (see Chap. 14). However, we can speak of acceleration *due* to gravity, thanks to the so-called *principle of equivalence*, the statement of Einstein's General Relativity according to which the acceleration which one experiences due to the gravitational attraction by a massive object is indistinguishable from an acceleration of the same magnitude which is produced by an appropriate force [2]. I found it very amusing and enlightening to discover, while reading a beautiful book by Jim Al-Khalili [3], that Einstein's principle of equivalence is the basis of operation of the roller coaster (or spaceship) simulators that can be found at the Luna Park or in a shopping mall. In a few words, the feeling to be glued to the seat when an airplane accelerates before taking off is the same (in principle) as that to be glued on the bed due to the gravitational attraction by the Earth although the latter is such a familiar feeling that we don't notice it. In the simulators, the lying position, which is equivalent to a forward acceleration of magnitude g (i.e., the acceleration to which we are subject due to gravity), together with the images on the screen, the wind from the fans, and other effects, creates the illusion of a roller coaster wagon or a spaceship which accelerate (with an acceleration equal to g).

3.2 Anesthesiologists, Intensivists, and Archimedes' Principle

The case which is usually of interest to anesthesiologists and intensivists is that where the "body" is represented by air bubbles. According to Archimedes' principle (and to common experience), air bubbles always tend to go upward in any liquid, including saline, drugs, and blood. A very trivial practical implication of this phenomenon pertains to how to handle a syringe if you realize that there is an air bubble inside and don't want to waste time, for example, to disconnect it from the tap along a venous line (Fig. 3.3); unless you do not need to inject the entire content of the syringe, just keep down the cone of the syringe!

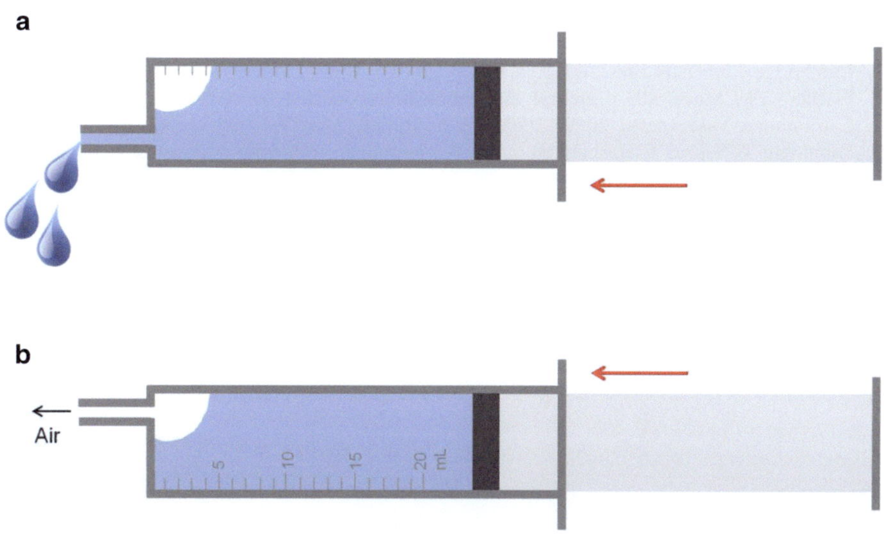

Fig. 3.3 Air bubble in a syringe. You can take advantage of Archimedes' principle in order to avoid injecting air to the patient. (**a**) cone down; (**b**) cone up

However, Archimedes' principle has more strictly clinical applications in cardiac surgery. During cardiac interventions which imply the opening of a main heart chamber, de-airing before aortic unclamping is needed in order to prevent air bubbles from reaching the brain and possibly causing ischemia. If de-airing is performed through a ventriculotomy, the anesthesiologist is asked to place the patient in the Trendelenburg position, so that the venting site is located above and air expulsion is favored [4]. Most noteworthy, Archimedes' principle helps cardiac anesthesiologists to prevent (or reduce) cerebral air embolism (AE) when air accidentally enters the circuits during cardiopulmonary bypass (CPB) or selective cerebral perfusion (SCP) in hypothermic circulatory arrest [5]. In these cases, in fact, the patient should be immediately placed in steep Trendelenburg position in order to protect the brain from AE.

Finally, AE with cerebral injury may also occur as a rare complication of central venous catheter (CVC) removal (especially with hemodialysis/hemofiltration catheters) [6]; although Trendelenburg position is not routinely used for the prevention of AE during CVC removal, it should be considered as an immediate intervention in case of suspected AE after this procedure.

We will come back on Trendelenburg position in Chap. 14, where we will see how physics (and trigonometry) can help anesthesiologists to use it safely.

References

1. Hess DR, Fink JB, Venkataraman ST, Kim IK, Myers TR, Tano BD. The history and physics of heliox. Respir Care. 2006;51:608–12.
2. Styer DF. Relativity for the questioning mind. Baltimore, MD: The Johns Hopkins University Press; 2011.
3. Al-Khalili J. Black holes, wormholes and time machines. 2nd ed. Boca Raton, FL: CRC Press; 2012.
4. Mackay JH, Arrowsmith JE, editors. Core topics in cardiac anesthesia. 2nd ed. Cambridge: Cambridge University Press; 2012. p. 178.
5. Fischer GW, Stone ME. Cerebral air embolism recognized by cerebral oximetry. Semin Cardiothorac Vasc Anesth. 2009;13:56–9.
6. Sahutoglu T, Sakaci T, Hasbal NB, et al. Air embolism following removal of hemodialysis catheter. Hemodial Int. 2017;21(1):29–34.

Dalton's Law and Fick's Law: Resorption Atelectasis, Membrane Oxygenators, and How an Air Bubble May Affect Blood Gas Analysis

4

Contents

Where's Physics

Daily life	The air we breathe Gas leaks Moisture on the walls of your home
Physics involved	Dalton's law Fick's law
Clinical practice	How an air bubble in the syringe affects blood gas analysis Absorption atelectasis Membrane oxygenators: • for cardiopulmonary bypass • for extracorporeal life support Physiology of respiration

4.1 Dalton's Law: When You Do the Math, It All Adds Up!

Dalton's law was enunciated by the English physicist of the nineteenth century from which also comes the name of the unified atomic mass unit (Da), as well as the term "daltonism," the color blindness of which he was suffering and that he described for

A. Pisano, *Physics for Anesthesiologists and Intensivists*, https://doi.org/10.1007/978-3-030-72047-6_4

the first time [1]. As mentioned in Chap. 2, it states that *the total pressure exerted by a mixture of non-reactive gases is equal to the sum of the partial pressures of individual gases*:

$$P_t = \sum_{i=1}^{m} P_i \quad \text{or} \quad P_t = P_1 + P_2 + \cdots + P_m \tag{4.1}$$

where $P_1, P_2, \ldots, P_m$ are the partial pressures of each component, P_t is the total pressure, and m is the number of gases in the mixture.

In other words, *each component of the mixture exerts a partial pressure equal to the pressure which it would exert if it occupied, alone, all the volume available* (see Fig. 4.1).

Thanks to this law, it can be easily demonstrated that the ideal gas law discussed in Chap. 2 (Eq. 2.1) also applies to a mixture of non-reactive gases, provided that the number of moles n is replaced with the sum of the numbers of moles of each component which are present in the mixture:

$$pV = \left(n_1 + n_2 + n_3 + \ldots\right) RT \tag{4.2}$$

where p is pressure, V is volume, R is the universal gas constant, T is the temperature (K), and n_1, n_2, n_3, etc. are the numbers of moles of each gas in the mixture.

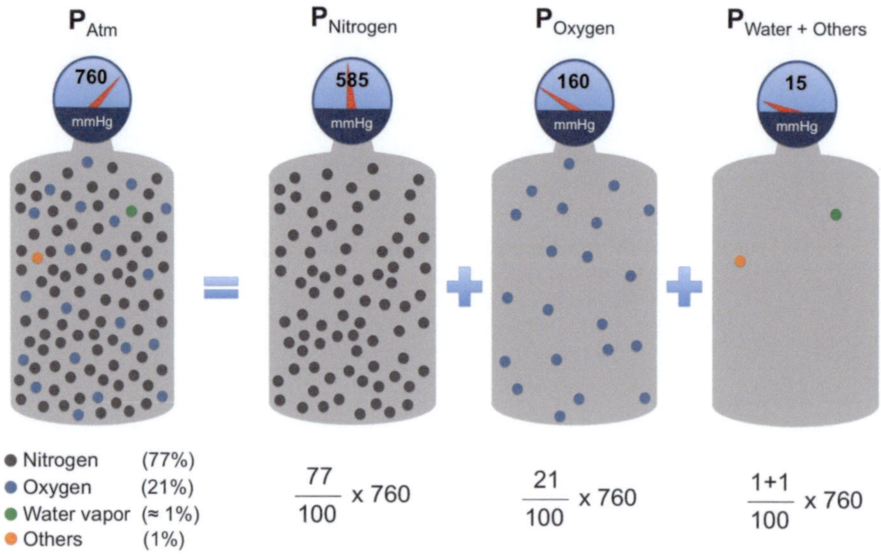

Fig. 4.1 Composition of atmospheric air and partial pressures of its components at sea level. It is clearly shown that the partial pressure of each component is equal to the pressure which it would exert if it occupied, alone, all the volume available (Dalton's law). P_{Atm} total atmospheric pressure

Consider the example of a mixture of two gases in which there are n_1 moles of one gas and n_2 moles of the other one. According to the universal gas law, we can write:

$$pV_1 = n_1 RT \tag{4.3}$$

and

$$pV_2 = n_2 RT \tag{4.4}$$

If the first gas occupied all the available volume ($V_1 + V_2$), its ideal gas law could be written as follows:

$$p_1 (V_1 + V_2) = n_1 RT \tag{4.5}$$

and, after replacing $n_1 RT$ with pV_1 (according to Eq. 4.3), we can calculate that its pressure p_1 would become:

$$p_1 = p \frac{V_1}{V_1 + V_2} \tag{4.6}$$

Similarly, if the second gas occupied, alone, all the available volume, its pressure would be:

$$p_2 = p \frac{V_2}{V_1 + V_2} \tag{4.7}$$

According to Dalton's law, the total pressure p of the mixture must be equal to $p_1 + p_2$. Therefore (using Eqs. 4.6 and 4.7):

$$p = p_1 + p_2 = \frac{pV_1}{V_1 + V_2} + \frac{pV_2}{V_1 + V_2} \tag{4.8}$$

and, after replacing pV_1 and pV_2 with $n_1 RT$ and $n_2 RT$, respectively (Eqs. 4.3 and 4.4), and $V_1 + V_2$ with the total volume V:

$$p = \frac{n_1 RT}{V} + \frac{n_2 RT}{V} \tag{4.9}$$

or

$$pV = (n_1 + n_2) RT \tag{4.10}$$

that is, indeed, what we wanted to prove.

Figure 4.1 shows the composition of atmospheric air and the calculation of the partial pressures of its main components according to Dalton's law. The air we breathe contains approximately 77% nitrogen, 21% oxygen, a variable amount (around 1%) of water vapor (which is called *humidity* and depends largely on temperature, as we will see in Chap. 15 when talking about "saturated vapor pressure"), while other gases such as carbon dioxide (CO_2), noble gases (mainly Argon), and very small quantities of other gases represent, overall, the remaining 1%.

As mentioned in Chap. 2, atmospheric pressure (P_{Atm}) at sea level is about 760 mmHg (or 101.3 kPa). According to Dalton's law, it must be:

$$P_{Atm} = p_{Nitrogen} + p_{Oxygen} + p_{Water} + p_{Others} = 760\,\text{mmHg} = 101.3\,\text{kPa} \quad (4.11)$$

It is therefore easy to calculate, for example, the partial pressure of nitrogen as:

$$P_{Nitrogen} = \frac{77}{100} \times 760\,\text{mmHg} \approx 585\,\text{mmHg} \approx 78\,\text{kPa} \quad (4.12)$$

To further clarify the meaning of Dalton's law, just think what happens when you stay in a room (with poor ventilation) where there is a gas leak. Even if the gas which is filling the room is not strictly "toxic," you will probably die if you cannot get out quickly. In fact, as the gas accumulates, its partial pressure increases and that of all the components of air, *including oxygen*, must decrease so that the total pressure remains (approximately) the same. Therefore, you would die (due to hypoxia) just to make everything fit according to Dalton's law! (However, a window and Fick's law could save your life, as discussed in the next section).

A similar (but less dramatic) example is the humidification and heating of air as it passes through the airways. When air reaches the alveoli, its partial pressure of water vapor becomes equal to the *saturated vapor pressure* of water at 37 °C (i.e., the highest partial pressure of water vapor allowable at this temperature, as we will see in Chap. 15), which is 5–6 times higher than the usual water vapor pressure in the atmosphere. As a consequence, according to Dalton's law, the partial pressures of other gases decrease. It is known, indeed, that alveolar partial pressure of oxygen (pO_2) is slightly lower than that of the "dry" inhaled air [2].

4.2 Down the Slope: Fick's Law

Two bodies (whatever their state of aggregation), which stay in contact for a long enough time, diffuse into one another until they form a homogeneous mixture. This even occurs between two solids, although "a long enough time" could be really too long! Conversely, the delicious smell coming from the kitchen (i.e., a vapor which spreads through air, or a "gas" which diffuses into another gas), the cocoa powder which mixes with my son's milk (i.e., a solid that diffuses into a liquid in which it is soluble), and oxygen which moves from inhaled air to blood through the alveoli (i.e., a gas that diffuses into a liquid) are common examples of *diffusion* which occurs rather quickly.

A substance always diffuses from where its concentration is higher to where it is lower (see Fig. 4.2). More precisely, diffusion occurs along a concentration gradient (G_C), that is the difference in concentration per unit distance or, more correctly, the slope of the graph of concentration against distance:

$$G_C = \frac{c_1 - c_2}{\Delta x} \quad \text{or} \quad G_C = \frac{dC}{dx} \quad (4.13)$$

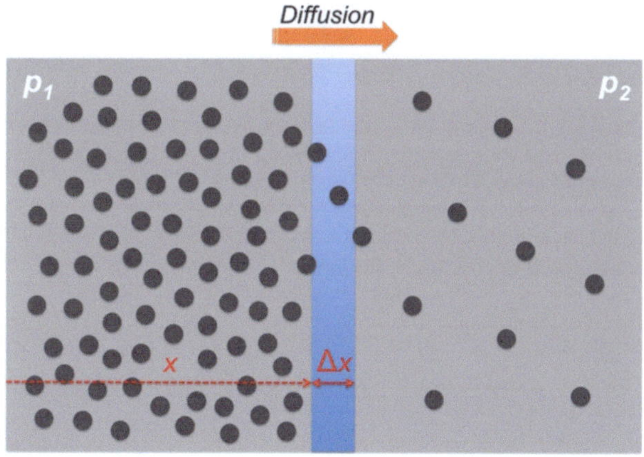

Fig. 4.2 According to Fick's law, a gas (whose molecules are represented by the little gray balls) diffuses trough air (whose molecules are not shown) from the left side of a box where it has the partial pressure p_1 to the right side of it where it has the partial pressure $p_2 < p_1$

where c_1 and c_2 are the concentrations of the substance at the point x and $x + \Delta x$, respectively, and dC/dx is the derivative of concentration with respect to distance (that is, as discussed in Chap. 1, the slope of the curve of concentration against distance).

In particular, *Fick's law* states that the *diffusion flux*, i.e., the amount q of substance (expressed in moles) which flows through a unit area A during a unit time interval t, is proportional to the concentration gradient G_C:

$$J = -D G_C \tag{4.14}$$

where J (= q/At) is the diffusion flux (mol m^{-2} s^{-1}), D is the so-called *diffusion coefficient* (or *diffusivity*) (see Box 4.1), and the minus sign means that, as mentioned, the flow is directed towards the areas with lower concentration (i.e., in the opposite direction to the gradient, or in the direction of a "negative" gradient).

Box 4.1: A Closer Look at "Diffusivity" of Gases

When talking about alveolar gas exchanges, everyone knows that CO_2 is much more *diffusible* than oxygen.

According to the kinetic theory of gases (see Chap. 2), the *average kinetic energy* $\bar{E}_K$ (see Box 2.2) is the same for all the molecules in a mixture of gases. Accordingly, if we consider a mixture of the gases 1 and 2, it must be:

$$\bar{E}_K = \frac{1}{2} m_1 \left(v_{RMS1}\right)^2 = \frac{1}{2} m_2 \left(v_{RMS2}\right)^2$$

where m_1 and m_2 are the molecular masses of the gases 1 and 2, respectively, and v_{RMS1} and v_{RMS2} are their root-mean-square speeds. The equation above can easily be rearranged as follows:

$$\frac{v_{\mathrm{RMS2}}}{v_{\mathrm{RMS1}}} = \sqrt{\frac{m_1}{m_2}}$$

or, remembering that the average velocity v is the distance Δx traveled during the time t (see Chap. 1):

$$\frac{t_1}{t_2} = \sqrt{\frac{m_1}{m_2}}$$

that is, the time that each component of the mixture needs to diffuse along a distance Δx is proportional to the square root of its molecular mass (*Graham's law of diffusion*, also known in some countries as Bunsen's law). According to this law, a gas with a higher molecular mass is less diffusible than one with a lower molecular mass.

If we remember the value of the molecular masses of oxygen (32 Da) and of CO_2 (44 Da), it is clear that *oxygen is more diffusible than* CO_2 ... and that there is something wrong! Really, there is no contradiction between Graham's law and the first sentence of this box, provided that you specify "when talking about alveolar gas exchanges."

Indeed, the *diffusion coefficient D* of Fick's law depends not only on the "solute" (i.e., the substance that diffuses), but also on the "solvent" (i.e., the substance in which the "solute" diffuses). When a gas diffuses into a liquid, the rate of diffusion is also directly proportional to the *solubility* of the gas in that liquid. Of course, CO_2 is much more soluble in blood than oxygen, so everything is explained! Finally, the solubility of a substance into another one also depends on temperature; in particular, the solubility of a gas into a liquid decreases with temperature (see Chap. 5), while the solubility of a solid into a liquid increases with temperature (that's why the cocoa powder mixes more easily with my son's milk if it is hot. Unfortunately, he wants it always cold!)

In the case where the substance which diffuses is a gas, the concentrations c_1 and c_2 can be replaced with the partial pressures p_1 and p_2, and it can be said that *diffusion occurs along a partial pressure gradient*. Figure 4.2 shows a shoe-box shaped container inside which a gas diffuses, through air, from the point x where it has the partial pressure p_1 to the point $x + \Delta x$ where it has the partial pressure p_2 (which is lower than p_1 by magic or because a partition between the two areas was just removed). In this example, Eq. (4.14) (Fick's law) can be written as follows:

$$J = \frac{q}{A\,t} = D\frac{p_1 - p_2}{\Delta x} \tag{4.15}$$

or

$$q = D\,A\,t\,\frac{p_1 - p_2}{\Delta x} \qquad (4.16)$$

that is, the amount q of gas which passes through the thickness Δx over the time t is directly proportional to the section of the container A, to the time t and to the difference in partial pressures of the gas across Δx, and inversely proportional to Δx itself.

The gas that has accumulated in a room due to a leaking pipe will rapidly flow out, thanks to the wide difference in its partial pressure between the room and the outside environment. Moreover, the higher the window size or the number of windows opened (i.e., the greater the area A through which the gas can pass) the faster the gas will diffuse away, just as described by Eq. (4.16). Of course, air (and therefore oxygen) will follow the reverse path.

As another example, the moisture (i.e., the partial pressure of water vapor) on the walls of your home also diffuses according to Fick's law [3].

4.2.1 Fick's Law and the Pathophysiology of Absorption Atelectasis

It is well-known that administration of high inspiratory oxygen fractions (FiO_2), such as for *preoxygenation* during general anesthesia induction, may promote the development of pulmonary atelectasis (especially if 100% oxygen is administered) due to rapid absorption of alveolar gas content through the alveolar-capillary membrane (*absorption atelectasis*) [4–7]. But why the alveoli tend to "empty" (and, therefore, to collapsing) if they contain mostly oxygen, while this does not occur when they contain mostly nitrogen (as usually)? It is often said that the reason is that oxygen is much more *diffusible* than nitrogen. If you have already read Box 4.1 (if not, do it before going on), you can easily guess that this statement needs some clarification. Of course, the diffusion of both oxygen and nitrogen through the alveolar-capillary membrane (ACM) occurs according to Fick's law (Eqs. 4.14–4.16); the amount of gas which diffuses through the ACM per unit time is proportional to the partial pressure gradient across the ACM and to a *diffusion coefficient* (also called, as mentioned, *diffusivity*). Due to its lower molecular mass as compared with oxygen (28 vs. 32 Da), nitrogen tends to diffuse, in general (e.g., through a hole in a container), a little faster than oxygen, according to the *Graham's law of diffusion*. On the other hand, as discussed in Box 4.1, the diffusion coefficient D of Fick's law (when considering respiratory gases crossing the ACM) also depends on the solubility of these gases in blood, which is higher for oxygen than for nitrogen. However, it is not so much the (slightly) higher diffusion coefficient of oxygen through the ACM as compared with that of nitrogen which explains the much greater absorption of oxygen, but rather the persistently high partial pressure *gradient* across the ACM due to the binding of oxygen with hemoglobin. In fact, unlike the partial pressure of nitrogen, the partial pressure of oxygen will never equalize between the two sides of the ACM since oxygen largely "disappears" (by binding to hemoglobin) after passing into the blood.

4.2.2 Fick's Law and Extracorporeal Support Technology: Membrane Oxygenators

Fick's law also regulates gas exchange within the *oxygenators* used for both cardiopulmonary bypass (CPB) during cardiac surgery and extracorporeal membrane oxygenation (ECMO) in patients with either acute respiratory failure or heart failure needing extracorporeal life support (ECLS). In membrane oxygenators (including "true" membrane oxygenators and the newest "hollow fiber" oxygenators), some kind of barrier, which is permeable to oxygen and carbon dioxide, separates the blood and gas phases, which move countercurrent; according to Eq. (4.16), the largest the surface area of this barrier, and the lowest its "thickness," the greater the gas exchange across it (which, of course, occurs depending on, and proportionally to, the partial pressure difference between the two "sides") [8]. A thorough discussion about the features and the operation of oxygenators is far beyond the scope of this book. Just as an example of how oxygenator technology has to contend with Fick's law, the earliest silicone membrane oxygenators needed much larger gas exchange surface areas (and, accordingly, larger sizes and priming volumes, with more hemodilution and more exposure to foreign materials) to provide adequate oxygenation and CO_2 removal as compared with the currently widely used "hollow fiber" oxygenators, such as the microporous oxygenators in polypropylene (PP) for CPB and the polymethylpentene (PMP) oxygenators for ECMO [8, 9]. In fact, thanks to their hollow fiber structure (something roughly equivalent to a reduction in membrane "thickness"), the latter have a much-improved gas exchange efficiency.

4.3 Air Bubbles and Blood Gas Analysis

We all know that we should not leave air bubbles in the syringe containing a blood sample for gas analysis because it can affect the results, especially pO_2 and partial pressure of carbon dioxide (pCO_2). Actually, even a gas bubble whose relative volume is 0.5 to 1% that of the blood in the collection device is a potential source of significant error [10, 11].

One might think, instinctively, that air bubbles reduce the pO_2, since we are accustomed to think of air as the most oxygen-poor mixture as compared to the inspiratory oxygen fractions (FiO_2) we administer usually. Instead, the literature reports an increased (i.e., overestimated) pO_2 due to air bubbles [11–13]. In order to understand why (and why this is not always true!), we have just to remember the two laws discussed above (particularly, Dalton's law).

Given a barometric pressure at sea level of 760 mmHg (101.3 kPa) and a concentration of oxygen of about 21%, the pO_2 of room air is, according to Dalton's law, approximately:

$$pO_2 = \frac{21}{100} \times 760\,\text{mmHg} = 159.6\,\text{mmHg} \approx 21.3\,\text{kPa} \tag{4.17}$$

Since a gas, according to Fick's law, diffuses from where its partial pressure is higher to where it is lower, oxygen will diffuse from bubbles to blood if patient's arterial pO_2 is lower than the pO_2 of air, and from blood to bubbles if patient's arterial pO_2 is higher than that of air. Accordingly, air bubbles will tend to increase (i.e., overestimate) the pO_2 value on arterial gas analysis when patient's arterial pO_2 is below 159.6 mmHg (or 21.3 kPa), as usually. However, if the patient has a higher arterial pO_2 (e.g., around 200 mmHg or 26.7 kPa), as may occur during oxygen supplementation, mechanical ventilation, CPB, or ECMO, with high FiO_2, then an air bubble in the syringe may lead to underestimation of pO_2 [14]. Moreover, it appears clear from Eq. (4.16) that the risk of a significant error in the determination of pO_2 is proportional to both the bubble size (the wider the "exchange surface" A, the higher the "rate of diffusion") and the delay between sampling and either analysis or bubble removal (the longer the contact time t between the bubble and blood, the higher the amount q of gas which diffuses from side to side) [12].

All this also applies to CO_2 (leaving out its different solubility in the blood as compared to oxygen). However, since the concentration of CO_2 in the air is very low (about 0.05%), the pCO_2 of air (approximately 0.38 mmHg or 0.05 kPa) is always much lower than that of the patient's arterial blood; accordingly, pCO_2 can only decrease (that is, be underestimated) due to the presence of an air bubble.

In conclusion, do not let air into the syringe, or at least remove it as soon as possible, when taking a blood sample for a gas analysis.

References

1. Fishman GA. John dalton: though in error, he still influenced our understanding of congenital color deficiency. Ophthalmic Genet. 2008;29:162–5.
2. Meyhoff CS, Staehr AK, Rasmussen LS. Rational use of oxygen in medical disease and anesthesia. Curr Opin Anaesthesiol. 2012;25(3):363–70.
3. Demars Y, Buck Y. Water vapor diffusion through insulated, masonry building walls. In: Lieff M, Trechsel HR, editors. Moisture migration in buildings: a symposium. ASTM International; 1982. p. 120.
4. Pisano A, Venditto M, Verniero L. Restrictive inspiratory oxygen fraction. In: Landoni G, Baiardo Redaelli M, Sartini C, Zangrillo A, Bellomo R, editors. Reducing mortality in critically ill patients. 2nd ed. Cham, Switzerland: Springer; 2021. p. 33–41.
5. Hedenstierna G, Meyhoff CS. Oxygen toxicity in major emergency surgery-anything new? Intensive Care Med. 2019;45(12):1802–5.
6. Bignami E, Saglietti F, Girombelli A, et al. Preoxygenation during induction of anesthesia in non-critically ill patients: a systematic review. J Clin Anesth. 2019;52:85–90.
7. Hedenstierna G, Edmark L. Mechanisms of atelectasis in the perioperative period. Best Pract Res Clin Anaesthesiol. 2010;24(2):157–69.
8. Hines MH. Principles of oxygenator function: gas exchange, heat transfer, and operation. In: Gravlee GP, Davis RF, Hammon JW, Kussman BD, editors. Cardiopulmonary bypass and mechanical support. Principles and practice. 4th ed. Alphen aan den Rijn, Netherlands: Wolters Kluwer; 2015. p. 95–108.
9. Khoshbin E, Roberts N, Harvey C, et al. Poly-methyl pentene oxygenators have improved gas exchange capability and reduced transfusion requirements in adult extracorporeal membrane oxygenation. ASAIO J. 2005;51(3):281–7.

10. Mueller RG, Lang GE, Beam JM. Bubbles in samples for blood gas determinations. A potential source of error. Am J Clin Pathol. 1976;65:242–9.
11. Lu JY, Kao JT, Chien TI, Lee TF, Tsai KS. Effects of air bubbles and tube transportation on blood oxygen tension in arterial blood gas analysis. J Formos Med Assoc. 2003;102:246–9.
12. Biswas CK, Ramos JM, Agroyannis B, Kerr DN. Blood gas analysis: effect of air bubbles in syringe and delay in estimation. Br Med J (Clin Res Ed). 1982;284:923–7.
13. Madiedo G, Sciacca R, Hause L. Air bubbles and temperature effect on blood gas analysis. J Clin Pathol. 1980;33:864–7.
14. Mueller RG, Lang GE. Blood gas analysis: effect of air bubbles in syringe and delay in estimation. Br Med J (Clin Res Ed). 1982;285:1659–60.

Cold, Sodas, and Blood Gas Analysis: Henry's Law

5

Contents

Where's Physics

Daily life	Sparkling water, carbonated drinks Flying corks (of Champagne) Bubbles in water which is not yet boiling
Physics involved	Henry's law
Clinical practice	Acid-base management during cardiopulmonary bypass (Alpha-stat vs. pH-stat) Pathophysiology and treatment of decompression sickness

5.1 The Physics in a Soda Bottle: Henry's Law

When you open a can of soda or a bottle of sparkling water and you see a lot of bubbles appear, you are observing the effects of *Henry's law*: the degree a gas dissolves into a fluid (i.e., the solubility of the gas in that fluid) increases proportionally to the increase in partial pressure above the fluid, according to the equation:

$$C_x = k_H \; P_x \tag{5.1}$$

where C_x and P_x are, respectively, the concentration and the partial pressure of the gas *x*, and k_H is the so-called Henry's law constant.

A. Pisano, *Physics for Anesthesiologists and Intensivists*,
https://doi.org/10.1007/978-3-030-72047-6_5

Carbonated water and fizzy soft drinks are produced by adding carbon dioxide (CO_2) to water under high pressure. Accordingly, before the can or the bottle are opened the gas above the drink (almost pure CO_2) is at higher pressure than atmospheric (Fig. 5.1). When the container is opened, some of this gas flees (producing the typical hiss) and its partial pressure above the liquid decreases; therefore, according to Henry's law, the solubility of CO_2 decreases, and part of the dissolved CO_2 comes out of solution as bubbles.

Moreover, have you ever noticed that warmer is the drink, larger are the bubbles? Or also, that on a hot day, drinks left in a glass become "flat" more quickly? This too is due to Henry's law. In fact, the Henry's constant is not at all constant (indeed, it should be called "Henry's coefficient"), but it depends on the gas, the solvent and the *temperature*, according to complex "Van't Hoff-like" equations that is not the case to report here [1–5]. In particular, the higher the temperature, the lower the value of k_H (and, hence, the solubility of the gas). Accordingly, when the drink is left in a glass, its characteristic acidulous taste (which is due to the CO_2 *dissolved* in the form of carbonic acid) tends to fade more quickly in a hot day since the drink warms faster.

Thermodynamically, it can be said that an increase in temperature translates into an increase in the kinetic energy of the gas molecules (see Chap. 2), resulting in a faster movement of them and in an easier breaking of intermolecular bonds, which enables molecules to escape from the solution and increases pressure. In other words, the partial pressure of a gas dissolved in a liquid is a measure of the kinetic energy of the gas molecules.

Fig. 5.1 The pressure of the gas (almost all CO_2) which is contained in a closed bottle of fizzy drink is slightly higher than atmospheric pressure (P_{Atm}). When the bottle is opened, this gas is released (producing the characteristic hiss) and replaced by air, whose partial pressure of CO_2 is very low (see Chap. 4, Sect. 4.3). According to Henry's law, the solubility of CO_2 in the liquid decreases and some CO_2 leaves the solution in the form of bubbles

Fig. 5.2 A pot of water on the stove. (**a**) The water is not yet boiling but you can see many tiny bubbles. These bubbles are made of gases (mainly oxygen and nitrogen) which leave the solution due to their reduced solubility into water as temperature rises (dependence on temperature of Henry's law constant). (**b**) The water is now boiling (the bubbles are made of water)

All this, applied to a bottle of Champagne, leads to another interesting revelation: the higher the temperature of the Champagne, the faster the cork will fly (and the greater will be the distance reached or the damage to breakable objects). In Champagne, CO_2 forms spontaneously due to fermentation of sugar. Accordingly, pressure is much higher than that in a fizzy soft drink (up to six times that of the atmosphere) [6] and a lot of energy is released when the bottle is opened. Most of this energy generates a sound shock wave (see Chap. 19) which produces the characteristic laud "pop," while a small part makes flying the cork [7]. Due to the temperature dependence of the Henry's coefficient, the higher the temperature, the lower the amount of CO_2 dissolved; this means a higher pressure in the bottle and, accordingly, a greater velocity (i.e., a higher kinetic energy) of the cork.

The relationship between Henry's coefficient and temperature also explains an event that can be observed by putting some water to boil: the little bubbles that appear before water starts to boil (Fig. 5.2). Someone might think that it is simply the slow beginning of boiling, but it is not. In fact, ebullition is, by definition, a tumultuous event (so, never slow) and the bubbles are made of water vapor. Instead, the "early" bubbles are formed by gases (such as oxygen) dissolved in water that come out of solution due to their reduced solubility with increasing temperature.

5.2 Acid-Base Management During Cardiopulmonary Bypass

One of the several things that alter the acid-base balance in a hypothermic patient [8–11] is what we have discussed above: the increase in the solubility of CO_2 into blood due to the lower temperature increases the concentration of dissolved CO_2 and, accordingly, reduces the partial pressure of CO_2 (pCO_2). This leads to respiratory alkalosis, which is masked by automated blood gas analysis (since it is performed at 37 °C), unless to correct values for the patient's temperature. Hence, there are two possible approaches to acid-base management during (hypothermic) cardiopulmonary bypass (CPB) or deep hypothermic circulatory arrest (DHCA) with selective cerebral perfusion (SCP): maintain normal values of pH and pCO_2 by adding CO_2 to the fresh oxygenator gas flow (pH-stat), or tolerate the respiratory alkalosis (Alpha-stat) [10, 12–14]. Although pH-stat theoretically improves cerebral perfusion (due to the higher pCO_2), this may lead to an increased risk of air embolism during CPB (we will see in Chap. 6 that it is again physics which explains why an air bubble can occlude a blood vessel). Indeed, literature seems now in favor of Alpha-stat in adult cardiac surgery, while pH-stat could be preferable during congenital procedures due to a more homogeneous brain cooling prior to circulatory arrest [14]. Beyond the choices and the different clinical settings, when addressing the "debate" on the best acid-base management strategy during CPB/DHCA, you may also think about boiling pots, carbonated drinks, flying corks, and Henry's law.

5.3 Pathophysiology and Treatment of Decompression Sickness

As mentioned in Chap. 2, decompression sickness (one of the two pathophysiological syndromes which are included in the more general term "decompression illness") occurs as a consequence of Henry's law [15, 16]. In fact, undersea the pressure increases by 1 atm ($\approx$760 mmHg or 101.3 kPa) for every 10 m of depth (according to Stevin's law, as discussed in Chap. 9). As depth increases, the amount of inert gases (mainly nitrogen) dissolved in the diver's blood increases according to Henry's law (the higher the ambient pressure and, according to Dalton's law, the partial pressure of the gas, the higher its solubility). If the diver ascents too fast, the amount of gas which leaves the solution due to decompression (according again to Henry's law) exceeds the rate of washout from tissues; this leads to supersaturation (and bubble formation). Recompression treatment in a hyperbaric chamber promotes the resolution of bubbles through several mechanisms (see Chap. 2) including, of course, Henry's law.

References

1. Fernández-Prini R, Alvarez JL, Harvey AH. Henry's constants and vapor–liquid distribution constants for gaseous solutes in H_2O and D_2O at high temperatures. J Phys Chem Ref Data. 2003;32:903–16.
2. Staudinger J, Roberts PV. A critical compilation of Henry's law constant temperature dependence relations for organic compounds in dilute aqueous solutions. Chemosphere. 2001;44:561–76.
3. Fernández-Prini R, Crovetto R. Evaluation of data on solubility of simple Apolar gases in light and heavy water at high temperature. J Phys Chem Ref Data. 1989;18:1231–43.
4. Majer V, Sedlbauer J, Bergin G. Henry's law constant and related coefficients for aqueous hydrocarbons, CO_2 and H_2S over a wide range of temperature and pressure. Fluid Phase Equilib. 2008;272:65–74.

5. Liger-Belair G. The physics and chemistry behind the bubbling properties of champagne and sparkling wines: a state-of-the-art review. J Agric Food Chem. 2005;53:2788–802.
6. Walker J. The flying circus of physics. 2nd ed. Wiley; 2007. p. 113–5.
7. Liger-Belair G, Bourget M, Cilindre C, Pron H, Polidori G. Champagne cork popping revisited through high-speed infrared imaging: the role of temperature. J Food Eng. 2013;116(1):78–85.
8. Kofstad J. Blood gases and hypothermia: some theoretical and practical considerations. Scand J Clin Lab Invest. 1996;56(Suppl 224):21–6.
9. Sanjay OP, Devnath A, Thejas BC. Arterial blood gas management during cardiopulmonary bypass. Indian J Clin Biochem. 2005;20:98–102.
10. Tallman RD Jr. Acid-base regulation, alpha-stat, and the emperor's new clothes. J Cardiothorac Vasc Anesth. 1997;11:282–8.
11. White FN. A comparative physiological approach to hypothermia. J Thorac Cardiovasc Surg. 1981;82:821–31.
12. Abdul Aziz KA, Meduoye A. Is pH-stat or alpha-stat the best technique to follow in patients undergoing deep hypothermic circulatory arrest? Interact Cardiovasc Thorac Surg. 2010;10:271–82.
13. Duebener LF, Hagino I, Sakamoto T, Mime LB, Stamm C, Zurakowski D, Schäfers HJ, Jonas RA. Effects of pH management during deep hypothermic bypass on cerebral microcirculation: alpha-stat versus pH-stat. Circulation. 2002;106:I103–8.
14. Grocott HP, Stafford-Smith M, Mora Mangano CT. Cardiopulmonary bypass management and organ protection. In: Kaplan's cardiac anesthesia: for cardiac and noncardiac surgery. 7th ed. Elsevier; 2016. p. 1111–61.
15. Vann RD, Butler FK, Mitchell SJ, Moon RE. Decompression illness. Lancet. 2011;377(9760):153–64.
16. Bennett MH, Lehm JP, Mitchell SJ, Wasiak J. Recompression and adjunctive therapy for decompression illness: a systematic review of randomized controlled trials. Anesth Analg. 2010;111(3):757–62.

Bubbles, Tracheal Tube Cuffs, and Reservoir Bags: Surface Tension and Laplace's Law

6

Contents

Where's Physics

Daily life	Inflate a party balloon; turn it into a sword or a cute bunny Soap bubbles and raindrops The strange behavior of party balloons, soap bubbles, and beer foam Coffee pod machines that jam
Physics involved	Surface tension Laplace's law
Clinical practice	Reservoir bag in anesthesia breathing systems Tracheal tube cuff and the "finger-pressure" technique Pathophysiology of air embolism, dilated cardiomyopathy, aortic aneurysm, and alveolar collapse (with due cautions) How an air bubble can clog a catheter

A. Pisano, *Physics for Anesthesiologists and Intensivists*,
https://doi.org/10.1007/978-3-030-72047-6_6

6.1 Physics in a Soap Bubble: Surface Tension and Laplace's Law

Laplace's law correlates the pressure difference between the inside and the outside of a bubble (or of a liquid drop) with its radius. Approximately, this relationship also applies to hollow objects with an elastic wall (including human organs) which contain a fluid (i.e., a liquid or a gas), provided that the wall is thin enough. Accordingly, examples abound. How does a soap bubble form? Why is inflating a balloon from your mouth easier as it becomes bigger? How can you turn a party balloon into a cute bunny? When the contents of two balloons, or of two bubbles (e.g., in the beer foam), come into contact, why does the smaller one become even smaller and the bigger one further enlarge? How does an aortic aneurysm grow? (And when to undergo surgery?) How can an air bubble stop blood flow in an arterial vessel (or even, for example, urine flow in a catheter)? Why do we need surfactant in pulmonary alveoli? How a (good) reservoir bag can prevent barotrauma during anesthesia? Laplace's law answers, at least in part (and in certain cases with some approximation), to all these questions. To deal with it, however, we have first to remind the concept of *surface tension*.

6.1.1 Also Liquids Care About Their "Appearance": Surface Tension

All molecules of which a liquid is composed attract one another due to (rather weak) *cohesive forces* acting within a distance of a few nanometers (1 nm = 10^{-9} m). A molecule within the liquid is surrounded by other molecules from each side. Accordingly, for reasons of symmetry, the *vector sum* (or *resultant*) (see Chap. 13) of all the attractive forces which act on it is equal to zero (Fig. 6.1). Conversely, on the molecules that are located near the free surface of the liquid acts a net force which is directed towards the inside of the liquid since such symmetry is lost. These forces are responsible for the so-called *surface tension,* which causes a liquid to always try to minimize its free surface and, accordingly, accounts for the roundish shape of raindrops, tears, and bubbles (as a sphere has the smallest possible surface area for a given volume). It is not entirely true, hence, that liquids have not a definite shape (as stated in Chap. 2) although the weak cohesive forces are easily exceeded by other forces (e.g., adhesive forces to the container) which often deprive liquids of their preferred shape!

In physics textbooks, surface tension is traditionally defined using the simple experimental device shown in Fig. 6.2. A thin liquid sheet is let form inside the space delimited by a U-shaped metallic frame and a thin metal wire which can slide along the frame without significant friction (see Chap. 14). Since the liquid surface tends to contract due to surface tension, a force F must be applied to keep the system in balance. Using a weighing scale, it can be shown that this force depends only on the frame width (i.e., on the length L of the edge of the liquid sheet). Considering that the liquid sheet has two sides, the surface tension γ can be defined as follows:

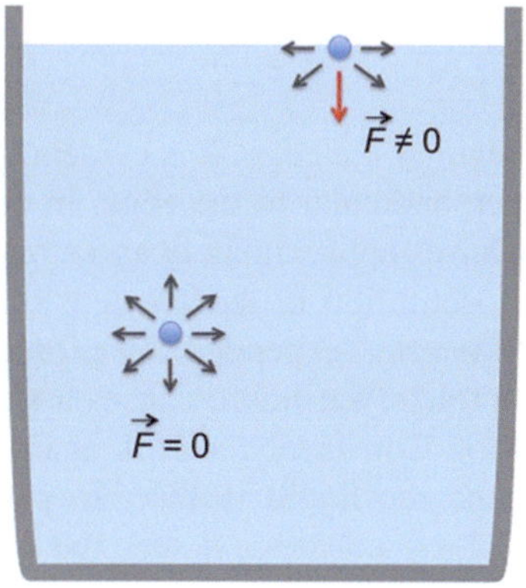

Fig. 6.1 Molecular basis of surface tension. Unlike a molecule within the liquid, which is equally attracted by other molecules from all directions, a molecule at the surface is mainly attracted towards the inside of the liquid since the kind (and the density) of molecules which are present above the surface is different. Indeed, the values of surface tension due to the contact between two substances (e.g., liquid-to-liquid or liquid-to-gas) depend, in general, on the characteristics of both the substances. *F* force

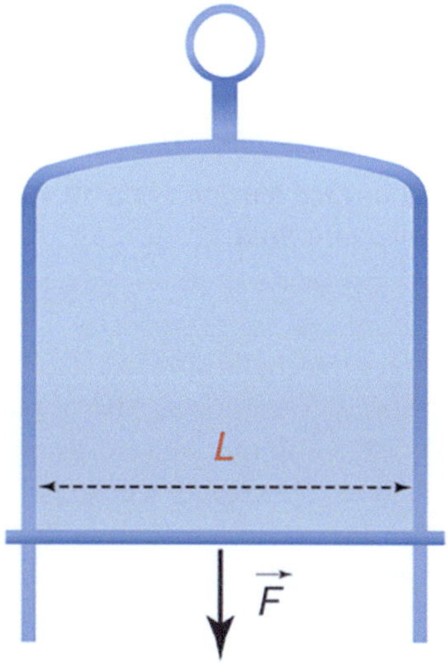

Fig. 6.2 Measuring surface tension (see text)

$$\gamma = \frac{F}{2L} \tag{6.1}$$

that is the *force per unit length of the edge*. It is evident that such a force is tangent to the liquid surface and perpendicular to the edge. In other words, due to surface tension, a liquid surface behaves approximately as a stretched elastic membrane.

An even more rigorous definition of surface tension relies on the concept of *mechanical work*, that is the *energy* "expended" by a force to set an object in motion (see Box 6.1). While it contracts, the liquid causes a shift of the metal wire and, accordingly, performs a work. Conversely, a work against surface tension must be performed in order to extend the liquid surface. In particular, if the metal wire moves (downward in Fig. 6.2) by a distance h, then the liquid surface will extend by an area:

$$S = 2Lh \tag{6.2}$$

where Lh is the area of a rectangle of base L and height h, multiplied by the two faces of the liquid sheet. As mentioned in Box 6.1, the work needed to get the shift h (and, accordingly, the extension $2S$ of the liquid surface) is:

$$W = F\,h \tag{6.3}$$

The work which is needed to increase by 1 m^2 the liquid surface is W/S, and it is easy to show that this work is equal to the surface tension itself (after substituting for W and S from Eqs. 6.3 and 6.2, respectively):

$$\frac{W}{S} = \frac{Fh}{2Lh} = \frac{F}{2L} = \gamma \tag{6.4}$$

Hence, it can be said that *surface tension is the work which is needed to extend the free surface of a liquid by a unit area.*

Box 6.1: Work Done by a Force (Mechanical Work)

As we will see in Chap. 14, a force is roughly "a push" or "a pull" which, when applied to an object, sets it in motion (unless opposed by other forces). By definition, the work W done by a force, whose point of application moves (since the object on which it's acting moves), is the *scalar product*, or *dot product*, of the force F and the displacement d caused by the force itself, that is:

$$W = F\,d\cos\theta$$

where θ is the angle between the directions of the force and the displacement. In other words, W is the product of the displacement and the magnitude of the component of the force along the displacement itself (see Chap. 13, Sect. 13.1). In the example of Fig. 6.2, the force F and the displacement h have the same direction: accordingly, $\theta = 0$ and, remembering that cos 0° = 1, the work W is simply $F \times h$ (Eq. 6.3).

Practically, work is the *energy* (see Chap. 7, Box 7.1) which is transferred to a body by a force acting on it. As mentioned, when you apply a (net) force to an object, it accelerates (see also Chap. 3, Box 3.1, and Chap. 14); this means that the kinetic energy of the object increases. Similarly, surface tension can also be defined as the *surface potential energy* of the liquid per unit area (in fact, due to surface tension, the liquid has "the potential" to do a work, which means that some energy is "stored" in it).

6.1.2 Laplace's Law

As stated above, liquids tend to assume a spherical shape due to surface tension. In fact, if the boundary surface of the liquid is a sphere (as in a raindrop or in a soap bubble), the forces due to surface tension (which are tangent to the surface) have a *resultant* which is perpendicular to the surface itself and, accordingly, lead to a pressure which "pushes" towards the center of the sphere. In order to understand how this pressure depends on surface tension and on the radius of the sphere, let's consider the soap bubble (i.e., a liquid "foil" of spherical shape) in Fig. 6.3. Imagine to inflate it a bit, so that its radius r becomes $r + \Delta r$. Remembering a little of mathematics (see Box 6.2), its external surface will increase by:

$$S = 4\pi\left(r+\Delta r\right)^2 - 4\pi r^2 = 4\pi\left(r^2+\Delta r^2+2r\Delta r\right) - 4\pi r^2$$
$$= 8\pi r\Delta r + 4\pi\left(\Delta r\right)^2 \approx 8\pi r\Delta r \qquad (6.5)$$

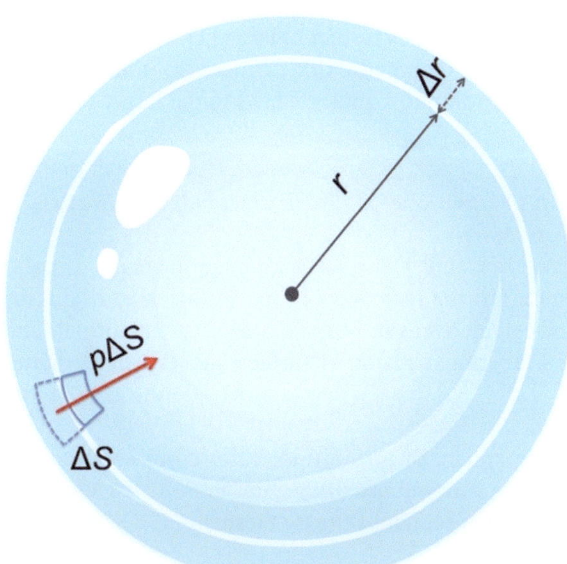

Fig. 6.3 Pressure (p) exerted perpendicularly to a portion ΔS of the (expanding) external surface of a soap bubble whose radius (r) increases by the value Δr

where $4\pi(\Delta r)^2$ can be neglected if Δr is very small as compared with r.

According to Eq. (6.4), the work to be done in order to obtain this size increase is:

$$W = S\gamma = 8\pi r \Delta r \gamma \tag{6.6}$$

At the same time, if p is the pressure due to surface tension and directed towards the center of the bubble, on each portion ΔS of the surface acts a force $p\Delta S$ (since pressure p is, by definition, force F per unit surface S, i.e., F/S), also directed towards the center (see Fig. 6.3). As the bubble size increases, the point of application of this force will move by a distance Δr. According to Eq. (6.3) (and to what described in Box 6.1), the work needed to extend each portion of the entire surface is the product of force (i.e., $p\Delta S$) and displacement Δr:

$$W_{\text{partial}} = p\,\Delta S\,\Delta r \tag{6.7}$$

To obtain the total work, we have to consider the sum of all the portions which form the surface, that is the surface area $4\pi r^2$ itself (see Box 6.2):

$$W_{\text{total}} = p 4\pi r^2\,\Delta r \tag{6.8}$$

From (6.6) and (6.8), is:

$$W = 8\pi r \Delta r \gamma = p 4\pi r^2 \Delta r \tag{6.9}$$

or (after solving for p):

$$p = \frac{2\gamma}{r} \tag{6.10}$$

This is a special case of the so-called *Laplace's law*. Hence, the higher the radius, the lower the pressure. This is why, in the beer foam, the smaller bubbles become even smaller and the bigger ones further enlarge [1]. In fact, according to Fick's law (as discussed in Chap. 4), carbon dioxide diffuses from the smaller bubbles (where pressure is higher, according to Laplace's law) to the larger ones.

Box 6.2: Two Little Memories of Mathematics (to Make Sense to Eq. 6.5)

Surface area of a sphere

As mentioned in Chap. 1, the surface area of a sphere of radius R is equal to $4\pi R^2$. Accordingly, the increase S in the surface area of the soap bubble in Fig. 6.3 can be calculated as its surface area after inflating it, when its radius has become $r + \Delta r$, minus its initial surface area (when its radius was r), i.e., $4\pi\,(r + \Delta r)^2 - 4\pi r^2$.

Square of a sum

In algebra, the square of a sum (also referred to as *square of a binomial* $A + B$) is the sum of the square of the first term A, the square of the second term B, and twice the product of the two terms:

$$(A+B)^2 = A^2 + B^2 + 2AB$$

Hence, in Eq. (6.5), $(r + \Delta r)^2$ becomes $r^2 + (\Delta r)^2 + 2\,r\Delta r$. In the next step, each of these three terms is multiplied by 4π, obtaining $4\pi r^2 + 4\pi(\Delta r)^2 + 8\pi r\Delta r$, and then the two terms $4\pi r^2$ and $-4\pi r^2$ cancel out.

6.2 Reservoir Bags and Tracheal Tube Cuffs

If a liquid bubble behaves as if it were surrounded by an elastic membrane, a hollow object which really has an elastic wall behaves (approximately) in the same way. When you inflate a balloon from your mouth, you make less effort as it becomes bigger; in fact, the pressure within the balloon decreases as its radius increases (actually, this happens only after the balloon has reached a certain size, since, unlike what occurs in liquid bubbles due to surface tension, the work needed to extend a "real" elastic membrane depends on its previous extension. At the beginning, the resistance to stretching leads to an increase in pressure [2]). If you have any doubts, you can easily perform, at your home, the "experiment" shown in Fig. 6.4.

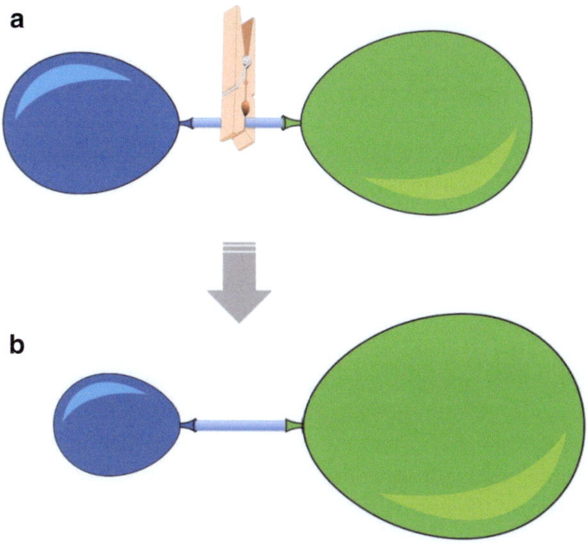

Fig. 6.4 In order to verify the effects of Laplace's law, you just need a couple of colored balloons, a straw, some adhesive tape, and a clothespin (or something like that). First put the clothespin on the straw, so as to prevent air flow between the two ends. Then, insert each end of the straw in a partially inflated balloon (and secure them with a bit of adhesive tape), making sure that one of the two balloons has been inflated more than the other. So, you built the system shown in (**a**). Now just remove the clothespin and see the smaller balloon becoming even smaller and the bigger one further enlarging (**b**). This occurs because the pressure is higher in the balloon with the smaller radius, according to Laplace's law

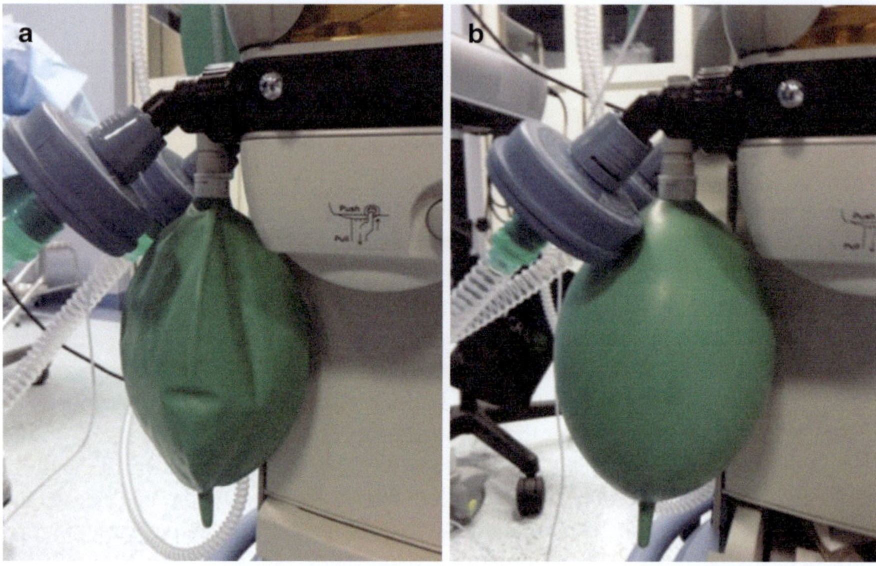

Fig. 6.5 A reservoir bag in an anesthesia machine. (**a**) With APL valve fully open, its radius and wall tension are small. (**b**) With APL valve closed, both wall tension and radius increase, so that the pressure inside remains constant (or *increases* slightly)

6.2.1 Some Unexpected Help from the Reservoir Bag

The reservoir bag in an anesthesia machine allows manual ventilation as well as a "visual" monitoring of spontaneous breathing. Moreover, thanks to Laplace's law, it can prevent barotrauma in case of malfunction or unintentional closing of the APL (adjustable pressure limiting) valve [3]. In fact, in the presence of an overflow or a flow obstruction in the breathing system, the radius of the reservoir bag increases (Fig. 6.5) and, according to Laplace's law, the pressure inside it decreases, thus preventing a dangerous rise in pressure in the entire breathing system and, consequently, in lungs. Accordingly, a reservoir bag which feels stiff should be replaced, since its *wall tension* (which we can define, similarly to surface tension γ, as the work required to extend the surface of an elastic membrane by a unit area) will be higher, for the same radius (or its radius will increase by a lesser extent for the same value of wall tension), thus providing a lower "pressure relief" (just have a look at Eq. 6.10).

6.2.2 Monitoring Tracheal Tube Cuff Pressure: We Cannot Trust Our Fingers

As mentioned in Chap. 2, the inflation pressure of the cuff of an endotracheal (or tracheotomy) tube should not exceed 30 cmH_2O ($\approx$22 mmHg) in order to prevent mucosal ischemia with possible consequent tracheal injury, including ulceration, tracheomalacia, stenosis, or tracheo-esophageal fistulae. On the other hand, pressures below 20 cmH_2O (≈15 mmHg) may allow microaspiration, thereby increasing the risk of ventilator-associated pneumonia (VAP) [4, 5].

Although pilot balloon palpation (the so-called *finger-pressure* technique) is still a widespread method to estimate intra-cuff pressure (especially in the operating room), it should not be used due to its poor accuracy [5, 6]. In particular, it has been shown that this technique often leads to significantly higher cuff pressures than recommended [7, 8]. In a recent investigation, Michlig asked 53 anesthesia practitioners to evaluate, by pilot balloon palpation, a cuff inflated to 88 mmHg (i.e., four times the maximum recommended value!) inside a syringe to mimic tracheal intubation (as shown in Fig. 6.6a): only 34% of them recognized overinflation [9].

A few years ago, I suggested that the poor accuracy of the "finger-pressure" method could be explained, at least in part, by Laplace's law [4]. In fact, although the pressure *p* inside the pilot balloon is practically the same that in the cuff (also according to *Pascal's principle*, as we will see in Chap. 9), what we really feel with

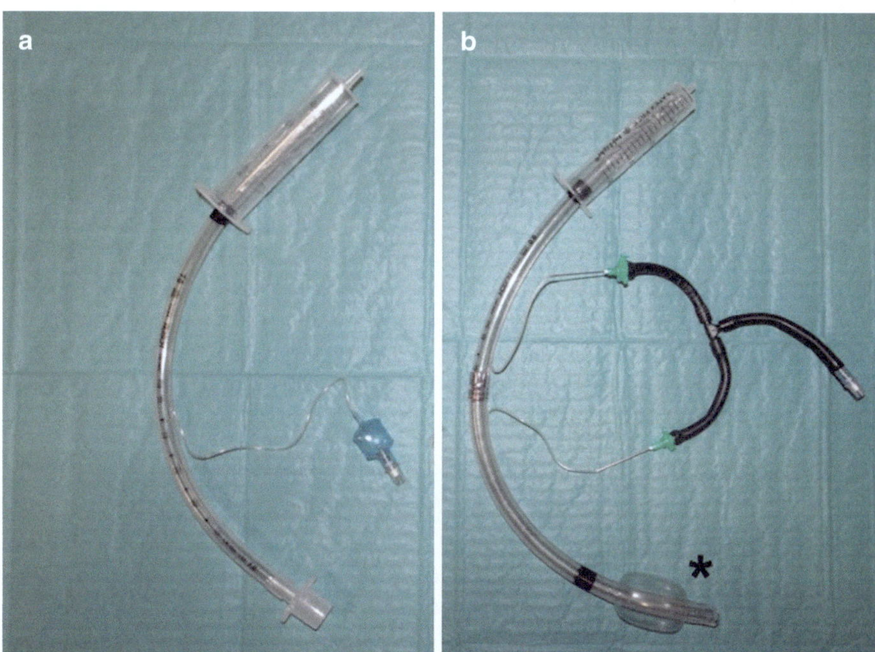

Fig. 6.6 The experimental set used to investigate the effect of the pilot balloon size on the reliability of the "finger-pressure" technique for intra-cuff pressure assessment (see text). (**a**) A normal endotracheal tube (Rüschelit®, Teleflex Medical, Wayne, PA) with its distal portion inserted into the barrel of a 20 mL syringe to simulate tracheal intubation. (**b**) A "modified tube" in which a Y-shaped tubing system equipped with a one-way valve connects the (identical) cuffs of two pieces of tube, so as to inflate one of them into a 20 mL syringe (as a simulated trachea) while using the other one (*) as a pilot balloon. Modified with permission from Pisano et al. [5]. Copyright © 2019 Springer Nature

our fingers is the *wall tension* T of the pilot balloon, that is (rearranging Eq. 6.10 and replacing surface tension γ with T):

$$T = \frac{pr}{2} \tag{6.11}$$

Since the radius r of the pilot balloon is smaller than that of the cuff, its wall tension is lower as compared with that of the cuff; accordingly, the pilot balloon will always seem "flabbier" than it is the cuff indeed. If we could touch the cuff directly, we would be probably more "accurate" in estimating its pressure. In order to verify this hypothesis, 62 anesthesia practitioners (41 experienced anesthesiologists and 21 residents) were asked to estimate the pressure of a cuff inflated to 88 mmHg into a simulated trachea as in the study by Michlig [9], but this time feeling both a usual pilot balloon and a modified "large" pilot balloon (identical to the cuff) which was obtained by connecting the cuffs of two different endotracheal tubes together, as shown in Fig. 6.6 [5]. After palpation of the usual pilot balloon, only 35% of participants (49% of experienced anesthesiologists and 10% of residents) recognized overinflation, as compared with 87% of participants (95% of experienced anesthesiologists and 71% of residents) after palpation of the modified pilot balloon. Moreover, 89% of participants (85% of experienced anesthesiologists and 95% of residents) stated that pressure was higher in the "large" pilot balloon than in the normal one (while they were exactly the same!). These results support the hypothesis that the pilot balloon size may affect the sensitivity of the "finger-pressure" technique (in particular, that the smaller radius of the pilot balloon as compared with the cuff favors underestimation of pressure due to the lower wall tension, according to Laplace's law). However, when the test was repeated with an intra-cuff pressure of 40 mmHg, only 32% of participants (51% of experienced anesthesiologists and none of the residents) recognized such a "moderate" overinflation after feeling the modified (larger) balloon. This confirms once again the poor reliability of the "finger-pressure" method in clinical practice [5].

6.3 Heart, Lungs, and Vessels (or Catheters)

Laplace's law (written in various ways and adapted to different situations) is often invoked as part of cardiovascular and pulmonary physiology, as well as of pathophysiology of several diseases such as dilated cardiomyopathy, aortic stenosis, and aortic aneurysms. Actually, things are often much more complex although the use of Laplace's law maintains its "educational" value. Finally, Laplace's law explains how an air bubble can stop circulation in an arterial vessel (or also urine flow in a urinary catheter).

In cardiovascular physiology, Laplace's law is often formulated in the following way:

$$\sigma = \frac{pr}{2h} \tag{6.12}$$

where σ is *wall stress* (i.e., tension acting over a cross-sectional area) and h is wall thickness (which appears due to the different definitions of wall "tension" and "stress"). Equation (6.12) applies to a spherical chamber (e.g., left ventricle, "sphere-shaped" aneurysms), while it becomes $\sigma = pr/h$ for a cylindrical vessel (that is, wall stress in a sphere for a given pressure is one half as compared with a cylinder with the same radius).

6.3.1 Left Ventricular Hypertrophy and Dilated Cardiomyopathy

A persistent increase in left ventricular afterload (such as in aortic stenosis) leads to an increased left ventricle (LV) pressure and, according to Laplace's law (Eq. 6.12), to a higher wall stress. Consequently, the myocardial oxygen consumption increases due to the greater energy which is needed in order to develop such increased tension. As a compensatory mechanism, LV hypertrophy develops (i.e., wall thickness increases) to reduce wall stress despite the increased LV pressure [10, 11].

In dilated cardiomyopathy, LV radius increases and, accordingly, a greater wall tension is needed to develop the same LV pressure. Heart failure occurs when wall tension to be developed becomes "too much." Of course, LV *is not* a sphere and, above all, wall stress is not homogeneously distributed along its thickness [10].

6.3.2 Aortic Aneurysm

An aortic aneurysm grows just like a party balloon (those of elongated shape, which are turned into cute bunnies by exploiting different radii of curvature, according to Laplace's law). When a weak spot develops in the aortic wall, the resulting expansion increases the diameter and, accordingly (Eq. 6.12), wall stress in that spot. As a consequence, the (weakened) vessel further enlarges [12]. Aneurysms often tend to assume a spherical shape in order to minimize wall stress (as stated, wall stress in a sphere is one half compared with a cylinder with the same radius). This, unfortunately, does not prevent wall rupture, which occurs when wall stress exceeds wall strength.

Surgeons mainly rely on diameter (and its progression) in order to estimate, according to Laplace's law, aortic wall stress and, hence, the risk of rupture and the indication for surgery [13, 14]. However, such an approach has been recently criticized, primarily because wall stress is not uniformly distributed within aortic wall, and other factors such as longitudinal stretch and vessel curvature should be considered [13–15].

6.3.3 Pulmonary alveoli Are Not a House of Cards

The belief that the smaller pulmonary alveoli can exist only thanks to the presence of surfactant (which reduces wall tension and, accordingly, internal pressure for a given radius) and that, in the absence of it, they would inexorably converge in a unique big alveolus (because of Laplace's law) should be reassessed [16]. Of course, surfactant plays a key role in preventing the collapse of small airways. However, alveoli also have a solid anatomy that supports them.

6.3.4 Air Embolism and Catheter Obstruction

In Chaps. 2 and 3, we discussed about air bubbles which could enter blood vessels during diving, cardiac surgery, and cardiopulmonary bypass (CPB), or even during removal of a central venous catheter (CVC). We know very well that this is hazardous since an air bubble may stop blood flow in a vessel, causing downstream ischemia (air embolism). However, one should be rather surprised by this... if (s)he does not remember Laplace's law! Indeed, it is *due to surface tension* (and according to Laplace's law) that a gas bubble carried by the blood flow can develop a pressure which is sufficient to counteract the pressure difference that allows blood to flow (see Chap. 8). In fact, under the "push" of blood, the two menisci of the bubble will deform as shown in Fig. 6.7. In order to understand why this occurs, we just need to remember the Hagen–Poiseuille equation (which is discussed in Chap. 8) and make a little clarification about Laplace's law. In Sect. 6.1.2, it is stated that, in a bubble, the pressure *due to surface tension* is proportional to the surface tension itself and inversely proportional to the radius of the bubble (Eq. 6.10). However, more generally, it is correct to say that the *pressure difference* across the bubble wall is, similarly, proportional to the surface tension and inversely proportional to the radius (as mentioned at the beginning of this chapter). According to Hagen–Poiseuille equation, a real (i.e., viscous) fluid needs a pressure difference between the two ends of a conduit to flow along it. Hence, blood will flow in the direction of the arrow in Fig. 6.7 only if blood pressure P_1 to the left of the bubble is higher than the pressure P_2 at its right. This means that the pressure difference Δp_1 across the left meniscus is lower than the pressure difference Δp_2 across the right one (since the pressure inside the bubble is the same for both sides). According to Laplace's law, a greater pressure difference across the meniscus implies a lower radius of curvature (as surface tension, which is due to the air-to-blood contact in both cases, is the same), and

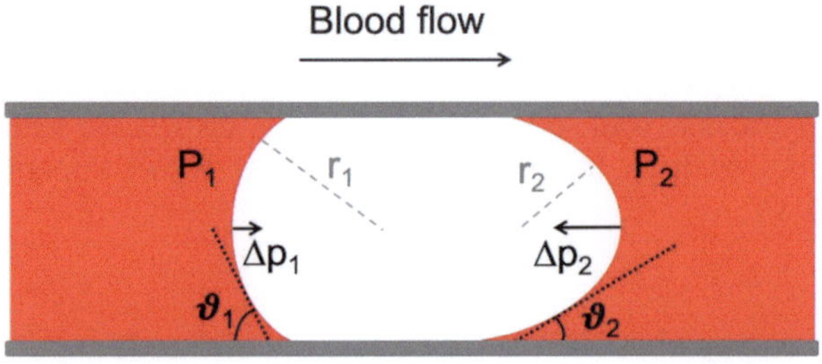

Fig. 6.7 Due to Laplace's law, an air bubble in a vessel deforms under the "push" of blood flow; as a consequence, surface tension can develop a pressure which is sufficient to stop blood circulation in the vessel (see text)

this is why the bubble assumes the asymmetrical shape shown in the figure. Using the definition of surface tension given in Sect. 6.1.1 and a bit of trigonometry (see Chap. 1), it is then possible to demonstrate that the component of the force due to surface tension along the direction of flow is higher for the meniscus with the smaller contact angle ϑ_2 (smaller radius of curvature r_2) than for the meniscus which tends to flatten out (higher radius of curvature r_1 and wider contact angle ϑ_1) (Fig. 6.7). In fact, since the force due to surface tension is tangential to the surface of the bubble, its *component* along the direction of flow is proportional to the cosine of the contact angle (in order to realize this, just have a look at Fig. 6.7 after having brushed up on the definition of vector component in Chap. 13, Sect. 13.1, or also in Box 6.1 in this chapter): if $\vartheta_2 < \vartheta_1$, then $\cos\vartheta_2 > \cos\vartheta_1$. The different magnitude of the two opposite forces (both perpendicular to the cross-sectional area of the vessel, but the most intense pointing against blood flow) leads to a *pressure* (that is, as mentioned above, force per unit area) which can be enough to offset the pressure difference which allows blood to flow.

Of course, all this also applies to urinary catheters (it can happen that "anuria" is simply due to an air bubble which obstructs the catheter) as well as to my coffee pod machine; sometimes, it feels like the water bottle housed inside is empty, but it is only an air bubble in the tube that sucks in the water!

References

1. Walker J. The flying circus of physics, 2nd ed. Wiley; 2007, pp. 115-116
2. Walker J. The flying circus of physics, 2nd ed. Wiley; 2007, pp. 207-208
3. Aston D, Rivers A, Dharmadasa A. Equipment in anaesthesia and critical care. Scion Publishing Limited. 2014;85
4. Pisano A. Pitfalls from physics: why we can't "feel" the tube cuff pressure with our fingers. Anesth Analg. 2017;124(4):1368.
5. Pisano A, Verniero L, Galdieri N, Corcione A. Assessing the correct inflation of the endotracheal tube cuff: a larger pilot balloon increases the sensitivity of the 'finger-pressure' technique, but it remains poorly reliable in clinical practice. J Clin Monit Comput. 2019;33(2):301–5.
6. Heffner JE, Hess D. Tracheostomy management in the chronically ventilated patient. Clin Chest Med. 2001;22(1):55–69.
7. Goyal R, Kumar G, Waghray MR. Endotracheal tube cuff pressure monitoring in peripheral hospitals. Med J Armed Forces India. 2006;62(3):243–5.
8. Liu J, Zhang X, Gong W, Li S, Wang F, Fu S, Zhang M, Hang Y. Correlations between controlled endotracheal tube cuff pressure and postprocedural complications: a multicenter study. Anesth Analg. 2010;111(5):1133–7.
9. Michlig SA. Anaesthetic staff cannot identify extremely high tracheal tube cuff pressures by palpation of the pilot balloon. Br J Anaesth. 2013;111(2):300–1.
10. Pagel PS. Cardiac physiology. In: Kaplan JA (Ed.) Kaplan's cardiac anesthesia, 6th ed. Elsevier Saunders; 2011, pp. 98-131
11. Sun LS, Schwarzenberger J. Cardiac physiology. In: Miller RD (Ed.) Miller's anesthesia, 6th ed. Elsevier Churchill Livingstone; 2005:723-741
12. Macura KJ, Corl FM, Fishman EK, Bluemke DA. Pathogenesis in acute aortic syndromes: aortic aneurysm leak and rupture and traumatic aortic transection. AJR Am J Roentgenol. 2003;181(2):303–7.

13. Sundt TM. Indications for aortic aneurysmectomy: too many variables and not enough equations? J Thorac Cardiovasc Surg. 2013;145(3 Suppl):S126–9.
14. Beller CJ, Gebhard MM, Karck M, Labrosse MR. Usefulness and limitations of computational models in aortic disease risk stratification. J Vasc Surg. 2010;52(6):1572–9.
15. Costanzo F, Brasseur JG. The invalidity of the Laplace law for biological vessels and of estimating elastic modulus from total stress vs. strain: a new practical method. Math Med Biol. 2015;32(1):1–37.
16. Prange HD. Laplace's law and the alveolus: a misconception of anatomy and a misapplication of physics. Adv Physiol Educ. 2003;27(1-4):34–40.

Part III

Fluids in Motion or at Rest: Masks, Tubes, Invasive Pressure Monitoring, and Hemodynamics

7 Continuity Equation and Bernoulli's Theorem: Airplanes, Venturi Masks, and Other Interesting Things (for Anesthesiologists and Intensivists)

Contents

Where's Physics

Daily life	How an airplane can fly Games with the hairdryer A simple experiment with party balloons Dangerous attractions (on the highway) Roofs blown away by the wind Watering the more distant areas of a garden Spraying people with Champagne
Physics involved	Bernoulli's theorem Continuity equation
Clinical practice	Venturi mask (Bernoulli's theorem and Continuity equation) Echocardiographic determination of valve area in aortic or mitral stenosis (Continuity equation) Echocardiographic estimation of pressure gradients (Bernoulli's theorem) Pathophysiology of dynamic left ventricular outflow tract obstruction due to systolic anterior motion (SAM) (Bernoulli's theorem and Continuity equation)

A. Pisano, *Physics for Anesthesiologists and Intensivists*,
https://doi.org/10.1007/978-3-030-72047-6_7

7.1 Garden Hoses and Echocardiographic Assessment of Heart Valve Stenosis: Continuity Equation

When you are watering a garden, you can reach the most distant areas by simply partially closing the hose with a thumb. In fact, the water speed increases when it flows through a decreased cross-sectional area. This occurs according to the so-called *continuity equation*, also known as *Leonardo's law* (after Leonardo da Vinci).

Figure 7.1 shows an ideal fluid flowing from the left to the right end of a tube segment with varying cross section. As an ideal fluid, we assume that it is incompressible (i.e., its density is constant), its *viscosity coefficient* is equal to zero (i.e., there are no internal forces which oppose the motion), and its flow is *laminar* (or *steady*), i.e., the speed of fluid at any point does not vary with time (see Chap. 8). Since the fluid is incompressible, the same volume ΔV which enters the tube at the left end during the time interval t must come out from the right end during the same time interval.

If a "particle" of the fluid moves by the distance Δx during the time interval t, then the volume ΔV which flows through the cross-sectional area A during the same time interval is:

$$\Delta V = A\,\Delta x \tag{7.1}$$

(remembering that the volume of a cylinder with base area B and height h is the product of B and h). Since, as recalled in Chaps. 1 (Sect. 1.4) and Chap. 4 (Box 4.1), the average speed v of a body is the distance Δx which the body travels during the time t ($v = \Delta x/t$ and, accordingly, $\Delta x = vt$), Eq. (7.1) can be written as follows:

$$\Delta V = A\,v\,t \tag{7.2}$$

where v is fluid speed.

If we apply Eq. (7.2) to the two ends of the tube segment of Fig. 7.1, we have:

$$\Delta V = A_1\,v_1\,t = A_2\,v_2\,t \tag{7.3}$$

that is:

$$\frac{\Delta V}{t} = A_1\,v_1 = A_2\,v_2 = \text{a constant} \tag{7.4}$$

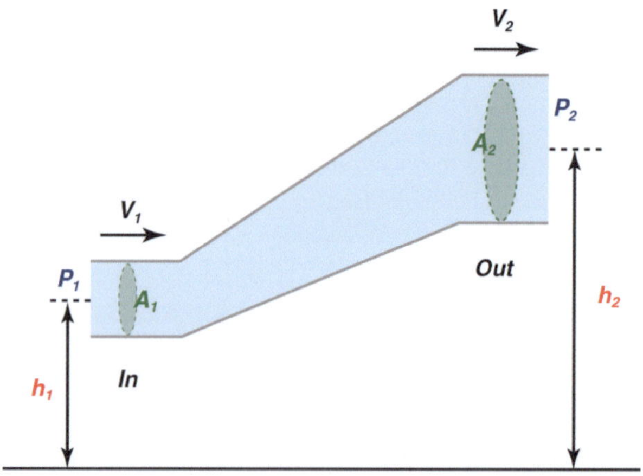

Fig. 7.1 A tube segment through which an ideal fluid flows from the input end (*In*) to the outlet end (*Out*). h_1, A_1, p_1, and v_1 and h_2, A_2, p_2, and v_2 are, respectively, the elevation, cross-sectional area, fluid pressure, and fluid speed at the two ends

where v_1, A_1, v_2, and A_2 are, respectively, the fluid speeds and the cross-sectional areas at the left end and at the right end of the tube, and $\Delta V/t$ is the so-called *volume flow rate* (that is, in a blood vessel, the cardiac output).

Hence, *the volume flow rate of an ideal fluid which flows in a tube is the same all along the tube*. Since the volume flow rate is the product of fluid speed and cross-sectional area, *the flow speed must increase where the cross-sectional area decreases* (continuity equation).

This law is used sometimes by Formula 1 or motorbike racers (who won the race) to drench with Champagne much people as possible (including those more distant, who can be reached by partially closing the mouth of the bottle with the thumb, as with a garden hose). Most importantly, it allows echocardiographers to calculate valve area as part of the evaluation of the severity of mitral or aortic stenosis [1, 2], particularly when planimetry is not possible (e.g., heavily calcified valve) or poorly reliable (e.g., associated valve regurgitation). As an example, according to continuity equation, the product of the aortic valve area (AVA) and the blood flow velocity through it, e.g., measured as aortic velocity-time integral VTI_{Ao} (see Chap. 20), must be equal to the product of the cross-sectional area of the left ventricular outflow tract (CSA_{LVOT}) and the VTI at this level (VTI_{LVOT}) [3]. Accordingly, AVA can be calculated as follows:

$$\mathrm{AVA} = \frac{\mathrm{CSA}_{\mathrm{LVOT}} \times \mathrm{VTI}_{\mathrm{LVOT}}}{\mathrm{VTI}_{\mathrm{Ao}}} \tag{7.5}$$

Leaving aside this stuff for echocardiographers (and for anesthesiologists and intensivists who are interested in echocardiography), we needed to know continuity equation before facing the Bernoulli's theorem, which is discussed in the next section. Subsequently, we will see how these two physical laws represent, together, the operation principle of the Venturi mask, and are also involved in the pathophysiology of dynamic left ventricular outflow tract obstruction due to systolic anterior motion (SAM), a possible cause of severe hemodynamic instability in both the operating room and the intensive care unit (ICU).

7.2 How Does an Airplane Fly? Bernoulli's Theorem

(Notice: if you really have never "digested" physics and/or mathematics, you can skip directly to Eq. (7.12). Conversely, the brave who want to fully understand can continue reading below). Let's consider again the tube segment shown in Fig. 7.1, in which an ideal fluid flows from the section A_1, which lies at a height h_1, to the section A_2 at a height h_2. As stated above, since the volume flow rate is constant, the fluid speed is inversely proportional to the cross-sectional area: accordingly, the speed v_2 of the fluid as it passes through the larger area A_2 is lower than the speed v_1 as it passes through the smaller area A_1. Finally, p_1 and p_2 are the pressures at the input and at the outlet, respectively.

Since we assumed that the fluid is nonviscous, its motion occurs without energy loss. In other words, the total mechanical energy of the system does not change (*principle of conservation of mechanical energy*) (see Box 7.1). One aspect of this

Box 7.1: To Know More: Energy and Conservation Laws

In physics, there are some quantities which, in an isolated system, do not change (or, as they say, "are conserved") as the system evolves over time. As we will see later in the book, *electric charge* (Chap. 11) and *linear momentum* (Chap. 14) are among them. Another conserved physical quantity is *angular momentum*, i.e., roughly, the product of rotation speed, mass, and "radius" of a rotating object (it is not the case to define it more rigorously here). As a practical and enjoyable example of what a conservation law may imply, an ice skater pirouettes faster (without adjunctive effort) when he/she brings his/her arms closer to the body. In fact, doing so he/she reduces his/her "radius." If we assume that the external torque acting on the skater is negligible, i.e., that he/she is "an isolated system," his/her angular momentum must remain constant, according to the law of conservation of angular momentum. Remembering that mass is another property which is conserved (at least in classical physics), the skater's rotation speed must increase when his/her "radius" decreases in order to keep the angular momentum constant.

One of the most important (and inviolable) physical principles is the *law of conservation of energy*: the total energy, i.e., the sum of mechanical energy (= kinetic energy + potential energy) and thermal energy (see Chap. 10) of an isolated system cannot change. If no dissipative forces (such as friction) act on an isolated system, which means that kinetic and potential energy are not irreversibly transformed into other forms of energy such as heat, then the mechanical energy of the system is also conserved (*law of conservation of mechanical energy*). This is why non-dissipative forces are also called "conservative" forces.

Defining "energy" is somewhat complex. Here we can say that *mechanical energy* is a "quantitative property" of an object which increases when some work (see Chap. 6, Box 6.1) is done on it, leading to an increase in the object's velocity (kinetic energy) or elevation (potential energy). For example, when we throw a ball in the air, its kinetic energy decreases as it slows down, while its potential energy increases (because it goes up while slowing down). When the ball reaches its maximum height (i.e., it stops), all its kinetic energy has transformed into potential energy, according to the law of conservation of mechanical energy (which applies if we consider air resistance, which is a non-conservative force, as negligible; in fact, the only force acting on the ball in this case is gravitational force, which is indeed conservative).

Just to know, according to modern physics (namely, from Einstein's Special Relativity onwards, see Chap. 21) neither mass nor energy are always conserved. What is really conserved is mass-energy; in a few words, mass and energy are "two sides of the same coin" and they can be transformed into each other (as the famous equation $E = mc^2$ states). Finally, there are other physical quantities which "are conserved," but they concern modern theoretical physics topics which are definitely not of interest to us here.

principle is the so-called *work-kinetic energy theorem*, which states that the work W that a force does on a body (see Chap. 6, Box 6.1) is equal to the change in kinetic energy (ΔE_k) of the body to which that force is applied:

$$W = \Delta E_k \tag{7.6}$$

Remembering the definition of kinetic energy (see Chap. 2, Box 2.2), for a "particle" of fluid with mass m whose speed changes from v_1 to v_2, is:

$$\Delta E_k = \frac{1}{2}mv_2^2 - \frac{1}{2}mv_1^2 \tag{7.7}$$

and, accordingly, the work-kinetic energy theorem can be written as follows:

$$W = \frac{1}{2}mv_2^2 - \frac{1}{2}mv_1^2 \tag{7.8}$$

were W is the work done by all the forces acting on the particle, which are, on the one hand, *gravity* (namely, the weight of the fluid particle) and, on the other hand, "surface" forces which push the fluid at the input level and are responsible for the pressure difference between p_1 and p_2.

The work done on the fluid particle by gravitational forces during the ascent from the height h_1 to the height h_2 is negative because the upward displacement $h_2 - h_1$ has an opposite direction to that of gravity. As discussed in Chap. 6 (Box 6.1), work is the scalar (dot) product of force F and displacement Δx, that is the product $F\Delta x$ multiplied by the cosine of the angle between the directions of F and Δx. In this case, the displacement is $\Delta x = h_2 - h_1$, the force is the weight of the particle, i.e., the product of its mass m and free-fall acceleration g (see Chap. 3, Box 3.1), and the cosine of the angle between force and displacement is equal to 1 since force and displacement have the same direction and, accordingly, the angle is zero (as mentioned in Chap. 1 or also in Box 6.1, cos 0° = 1). Therefore, the "gravitational" work W_g is:

$$W_g = -m\,g\left(h_2 - h_1\right) = m\,g\,h_1 - m\,g\,h_2 \tag{7.9}$$

Regarding the "surface" forces, imagine that the pressure difference p_1 - p_2 makes the fluid particle move by the tract Δx. Then, the work W_p done due to this pressure difference is:

$$W_p = F\,\Delta x = \left(p_1 - p_2\right)A\,\Delta x = \left(p_1 - p_2\right)A\frac{V}{A} = \left(p_1 - p_2\right)V = \left(p_1 - p_2\right)\frac{m}{d} \tag{7.10}$$

where F has been replaced with $(p_1 - p_2)A$ remembering that, by definition, pressure is force per unit area ($p = F/A$ and, accordingly, $F = pA$); Δx has been replaced with V/A remembering that the volume V of a cylinder of base area A and height Δx is $V = A\Delta x$ (and, accordingly, is $\Delta x = V/A$); and, finally, V has been replaced with m/d

remembering that, by definition, density is mass per unit volume ($d = m/V$ and, hence, $V = m/d$).

Substituting by Eqs. (7.8), (7.9), and (7.10), the work-kinetic energy theorem (Eq. 7.6) can then be written as follows:

$$W = W_g + W_p = m\,g\,h_1 - m\,g\,h_2 + (p_1 - p_2)\frac{m}{d} = \frac{1}{2}mv_2^2 - \frac{1}{2}mv_1^2 \quad (7.11)$$

or, after some simple rearrangements (divide by m and multiply by d all the terms):

$$d\,g\,h_1 + p_1 + \frac{1}{2}dv_1^2 = d\,g\,h_2 + p_2 + \frac{1}{2}dv_2^2 = \text{a constant} \quad (7.12)$$

that is the so-called *Bernoulli's equation* (or Bernoulli's theorem), where h_1, p_1, and v_1 are the elevation, pressure and speed of a fluid entering a tube at one end, h_2, p_2, and v_2 the corresponding variables for the fluid emerging at the other end (Fig. 7.1), d is the fluid density, and g is the free-fall acceleration.

7.2.1 And Now...Let's Fly This Airplane!

Bernoulli's theorem tells us something very important when considering the flow of a fluid that travels along a horizontal streamline, so that its elevation does not change. In this case, we have $h_1 = h_2$ and the previous equation becomes:

$$p_1 + \frac{1}{2}dv_1^2 = p_2 + \frac{1}{2}dv_2^2 = \text{a constant} \quad (7.13)$$

This means that *where the speed is higher the pressure is lower, and vice versa*. Bernoulli's theorem remains valid even for a fictitious *tube of flow*, i.e., in the absence of a real conduit. You can easily verify this at your home using two party balloons and a straw, as shown in Fig. 7.2. Another "homemade experiment" to show the effects of Bernoulli's theorem can be made with a hairdryer and a sheet of paper. Place the sheet on a table, turn on the hairdryer, and approach it to the middle of the sheet, maintaining the air jet parallel to the sheet. At this point, air speed above the sheet will be much higher than below it and, according to Eq. (7.13), the pressure below the sheet will be higher than the pressure above it. Hence, you should see the sheet that rises up and "adheres" to the hairdryer. Really, this is not very easy to perform; the sheet will fly away many times until you don't find the right "technique." However, if you succeed, besides seeing with your own eyes the Bernoulli's theorem at work, you will be very close to discover how an airplane can fly. In fact, with an extreme simplification, it can be said that the particular airfoil shape of the wings (the top more convex than the bottom) yields a higher speed of air above wings and a lower speed below them, due to the greater distance which, during the same time, is traveled by air near the upper surface of the wings as

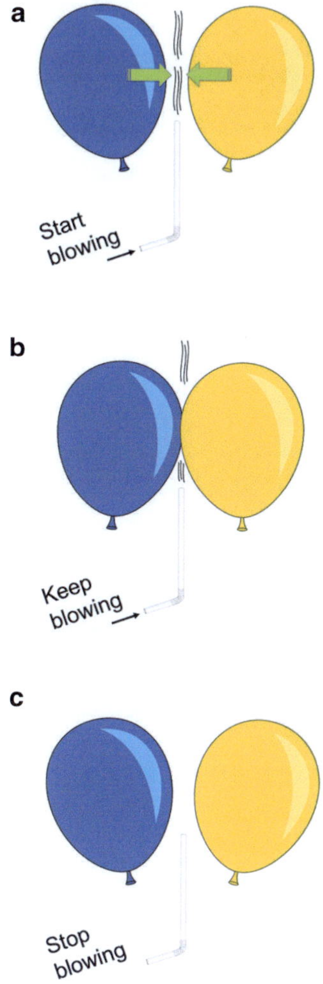

Fig. 7.2 A simple experiment to show the effects of Bernoulli's theorem. Ask someone to keep two party balloons at a distance of 1–2 cm each other. Using a straw, blow between the two balloons; they will tend to get closer to each other (**a**), until they join if you blow hard enough (**b**). This is due to the reduction in pressure between the balloons (as compared to the atmospheric pressure all around) as air speed increases, according to Bernoulli's theorem. When you stop blowing, the balloons separate again (**c**)

compared to the lower surface (Fig. 7.3). According to Bernoulli's theorem, this implies a lower pressure on the top surface of the wings and a higher pressure on the bottom surface, hence generating the lift in the upward direction [4]. The airplane takes off when its speed is sufficient so that such upward lift overcomes its weight.

Unfortunately, similarly to how it allows airplanes to fly, Bernoulli's theorem can also make the roofs of houses fly! In fact, the strong wind blowing over the roof during a storm (a hurricane is usually needed) generates a pressure difference between above and below the roof which pushes upward and can contribute to blow away the roof. Of course, different building techniques allow to minimize this risk [5].

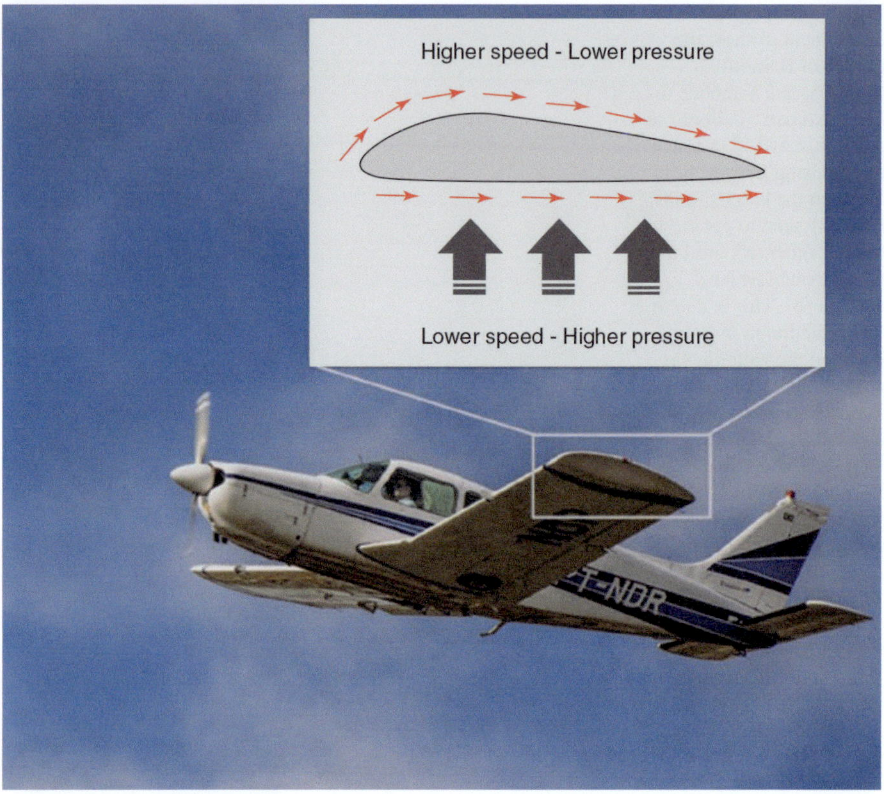

Fig. 7.3 How an airplane can fly. Since the top of the wing is more curved than its bottom, the air flow (red arrows) has a higher speed near the upper surface as compared to the lower surface. This accounts, according to Bernoulli's theorem, for the pressure difference which generates the so-called lift (upward arrows)

7.2.2 Bernoulli's Theorem and Echocardiography

A simplified version of Bernoulli's equation allows echocardiographers to estimate maximum pressure gradients (e.g., across a heart valve) from the maximum jet velocity measured by Doppler analysis (see Chap. 20) [6].

Solving for $p_1 - p_2$, Eq. (7.13) becomes:

$$p_1 - p_2 = \frac{1}{2}dv_2^2 - \frac{1}{2}dv_1^2 \tag{7.14}$$

Assuming that the velocity v_1 proximal to the site of interest is negligible (i.e., is much lower than v_2) [7, 8], Eq. (7.14) can be written as:

$$p_1 - p_2 = \Delta p = \frac{1}{2}dv_2^2 \tag{7.15}$$

or, after replacing d with an appropriate coefficient that includes an approximated value for blood density and takes into account the necessary unit conversions:

$$\Delta p = 4v_2^2 \tag{7.16}$$

which is the *simplified Bernoulli's equation* usually reported in echocardiography textbooks. According to Eq. (7.16), the (maximum) pressure "gradient" Δp across a heart valve can be calculated from the peak blood velocity across it (measured by Doppler analysis, as discussed in Chap. 20). For example, the maximum pressure gradient across a stenotic aortic valve is measured as part of the assessment of stenosis severity [6], while measurement of the maximum pressure gradient across the tricuspid valve (in the presence of some degree of tricuspid regurgitation) allows estimation of peak right ventricle pressure (and, hence, of systolic pulmonary artery pressure) by adding the value of central venous pressure (CVP) to the gradient [9].

More details about these topics are beyond the scope of this book. However, you will find something more on echocardiography in Chap. 20, in which we will talk about ultrasounds and Doppler effect.

7.3 Continuity Equation and Bernoulli's Theorem Work Together in a Venturi mask

Even if someone in the past didn't agree [10], it is now generally accepted that also the way the Venturi mask works relies on Bernoulli's theorem [11, 12]. In a Venturi mask (Figs. 7.4 and 7.5), 100% oxygen flows through tubing at a specified flow rate.

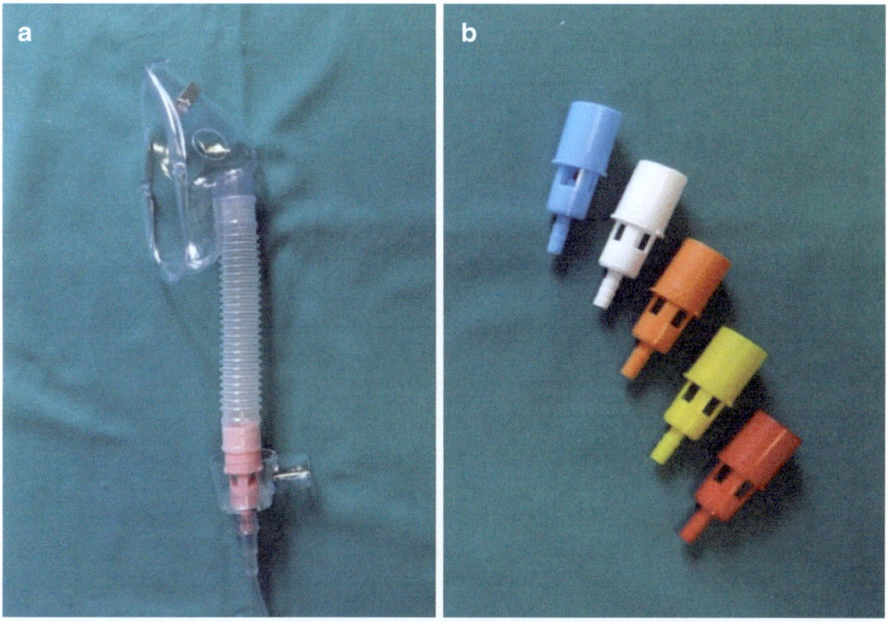

Fig. 7.4 (**a**) A Venturi mask assembled with a 50% (pink) Venturi valve. (**b**) Different Venturi valves allow the administration of different inspired oxygen fractions: blue, 24%; white, 28%; orange, 31%; yellow, 35%; red, 40%

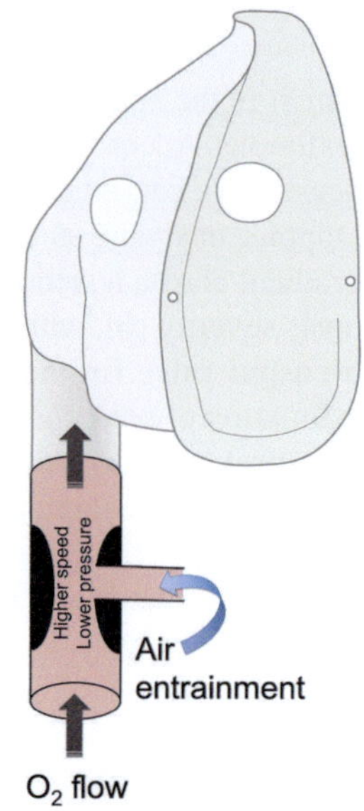

Fig. 7.5 Schematic representation and principle of operation of a Venturi valve (see text). The interface through which the patient breathes is also shown. O_2 Oxygen

At the level of the so-called Venturi valve (a schematic representation of which is provided in Fig. 7.5), the pipe narrows (i.e., its cross-sectional area decreases) and there is a side port. According to continuity equation, the speed of oxygen increases as it flows through such narrowing and, according to Bernoulli's theorem, this leads to a drop in pressure (at subatmospheric values) that allows entrainment of atmospheric air from the side port. The reduction of pressure in a fluid which flows through a constriction along a pipe is usually referred to as "Venturi effect," after Giovanni Battista Venturi who described this phenomenon in the late eighteenth century (hence also the name "Venturi mask"). However, now we know that there are two more general physical laws behind this "effect." Returning to the Venturi mask, different preformed Venturi valves, of different colors (Fig. 7.4b), allow the administration of *predictable* inspiratory oxygen fractions (FiO_2) (the greater the narrowing, the lower the FiO_2), provided that the oxygen flow rate indicated on each Venturi valve is used.

7.4 Continuity Equation and Bernoulli's Theorem Work Together Again: The Pathophysiology of Systolic Anterior Motion

Maybe you happened to experience for yourself the combined effects of continuity equation and Bernoulli's theorem (or, if you prefer, the Venturi effect) while overtaking a truck on the highway with your car at high speed. If you get too close to the truck when overtaking it, you will notice that the car is (dangerously) attracted towards the truck; in fact, as air flows through a narrowing (i.e., the space between your car and the truck), its speed increases in order to keep volume flow rate constant (continuity equation) and, according to Bernoulli's theorem, air pressure

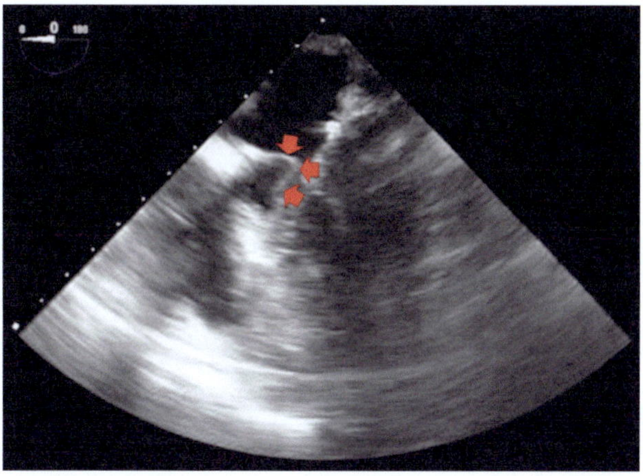

Fig. 7.6 Systolic anterior motion (transesophageal echocardiography). The anterior mitral leaflet (red arrows) is attracted towards the left ventricular outflow tract (LVOT) during systole, causing (dynamic) LVOT obstruction and mitral regurgitation

between the two vehicles decreases as compared with the atmospheric pressure all around, leading to the attraction of the car towards the truck (not the contrary because of the higher mass of the truck). A similar mechanism (also think again of the balloons in Fig. 7.2) explains, at least in part, the "attraction" of the anterior mitral leaflet towards the left ventricular outflow tract (LVOT) during systole which can occur in some circumstances (Fig. 7.6) [13]. This pathological condition is known as *systolic anterior motion* (SAM) and causes, at the same time, mitral regurgitation and (dynamic) LVOT obstruction, often leading to severe hemodynamic instability. Predisposing and precipitating factors include all conditions which narrow the LVOT or anyway favor the increase in blood speed within it, such as hypertrophic cardiomyopathy, hypovolemia, tachycardia (which impairs diastolic filling), increased contractility, and reduced left ventricular afterload [13, 14]. For example, neuraxial anesthesia (causing significant vasoplegia) can induce SAM in a patient with severe septal hypertrophy (which, in fact, represents a contraindication to neuraxial anesthesia), and adding inotropic drugs to treat the consequent severe hemodynamic instability usually only worsens the situation. SAM may also occur in the absence of structural predisposing factors (myocardial hypertrophy), e.g., in sepsis, due to reduced left ventricular afterload, tachycardia, and increased myocardial contractility [13]. Considering the mechanism of SAM-induced dynamic LVOT obstruction, its treatment may include fluids, vasopressors, ß-blockers, and discontinuation of inotropic drugs.

References

1. Messika-Zeitoun D, Iung B, Brochet E, Himbert D, Serfaty JM, Laissy JP, Vahanian A. Evaluation of mitral stenosis in 2008. Arch Cardiovasc Dis. 2008;101:653–63.
2. Otto CM. Valvular aortic stenosis: disease severity and timing of intervention. J Am Coll Cardiol. 2006;47:2141–51.

3. Otto CM. Textbook of clinical echocardiography, 5th ed. Elsevier Saunders; 2013, pp. 280–284.
4. Anderson J. Introduction to flight, 7th ed. McGraw-Hill; 2011, pp. 352–362.
5. Singh J, Roy AK. Effects of roof slope and wind direction on wind pressure distribution on the roof of a square plan pyramidal low-rise building using CFD simulation. Int J Adv Struct Eng. 2019;11:231–54.
6. Otto CM. Textbook of clinical echocardiography, 5th ed. Elsevier Saunders; 2013, p. 278.
7. Heys JJ, Holyoak N, Calleja AM, et al. Revisiting the simplified Bernoulli equation. Open Biomed Eng J. 2010;4:123–8.
8. Maslow AD, Bert AA. Echocardiographic evaluation of prosthetic valves. In: Oxorn DC, editor. Intraoperative echocardiography. Philadelphia, PA: Elsevier Saunders; 2012. p. 95–130.
9. Schroeder RA, Mark JB, Barbeito A. Right-sided valvular disease. In: Oxorn DC, editor. Intraoperative echocardiography. Philadelphia, PA: Elsevier Saunders; 2012. p. 77–94.
10. Scacci R. Air entrainment masks: jet mixing is how they work; the Bernoulli and Venturi principles are how they don't. Resp Care. 1979;24:928–31.
11. Soto-Ruiz KM, Peacock WF, Varon J. The men and history behind the Venturi mask. Resuscitation. 2011;82:244–6.
12. Adcock CJ, Dawson JS. The Venturi mask: more than moulded plastic. Br J Hosp Med (Lond). 2007;68:M28–9.
13. Dugar S, Latifi M, Moghekar A, Duggal A. All shock states are not the same. Systolic anterior motion of mitral valve causing left ventricular outflow tract obstruction in septic shock. Ann Am Thorac Soc. 2016;13(10):1851–5.
14. Luckner G, Margreiter J, Jochberger S, et al. Systolic anterior motion of the mitral valve with left ventricular outflow tract obstruction: three cases of acute perioperative hypotension in noncardiac surgery. Anesth Analg. 2005;100(6):1594–8.

From Tubes and Catheters to the Basis of Hemodynamics: Viscosity and Hagen–Poiseuille Equation

8

Contents

Where's Physics

Daily life	"True" liquids and their motion (why we really need a heart) Pour ketchup and mayonnaise Walking on water
Physics involved	Viscosity Hagen–Poiseuille equation Non-Newtonian fluids Laminar or Turbulent flow (Reynolds number)
Clinical practice	Endotracheal/tracheotomy tubes and weaning from mechanical ventilation Work of breathing (WOB) Peripheral vs. central venous catheter "Resistance" of an epidural or pulmonary artery catheter Understanding hemodynamic monitoring and management: the relationship among blood pressure, cardiac output, and vascular resistances Applying ultrasound gel on the probe Cannulae for Extracorporeal Membrane Oxygenation (ECMO) Sounds listened to with the stethoscope

A. Pisano, *Physics for Anesthesiologists and Intensivists*,
https://doi.org/10.1007/978-3-030-72047-6_8

8.1 Real Fluids Flow in a Different Way: Viscosity and Hagen–Poiseuille Equation

According to Bernoulli's theorem (see Chap. 7, Eq. 7.12), the pressure of an ideal fluid which flows through a horizontal tube with constant cross-sectional area is the same all along the tube; in fact, the elevation does not change and, due to continuity equation (Chap. 7, Eq. 7.4), also the fluid speed is constant (since the cross-sectional area is constant). Conversely, if we measure pressure along a tube in which a *real* fluid flows, we will find that it gradually decreases down the direction of flow. This occurs because, unlike ideal fluids (which we assumed to be nonviscous), real fluids are characterized by *viscosity*, i.e., "internal friction" forces which oppose to motion and, accordingly, cause energy loss during it.

Let's briefly define viscosity, so as to be able to discuss the law which describes the flow of a real fluid through a cylindrical tube and its many implications in anesthesiology and critical care medicine practice.

8.1.1 Viscosity

Figure 8.1 shows a plate which moves on the free surface of a liquid. As we will see in Chap. 14, according to *Newton's first law* (also referred to as "law of inertia"), no (net) forces act on a body which is moving with *constant* speed on a frictionless surface. Accordingly, if a force F is needed in order to maintain the plate in uniform rectilinear motion (i.e., at a constant speed), it means that the liquid surface is not frictionless, and that such a force F must be equal to the friction forces; in fact, if it were lower, the plate could not move (since friction forces are those which oppose to motion), while if it were higher, its motion would be accelerated according to Newton's second law (see Chap. 3, Box 3.1, and Chap. 14). Practically, the liquid layer which is in contact with the surface S of the plate will move with the same

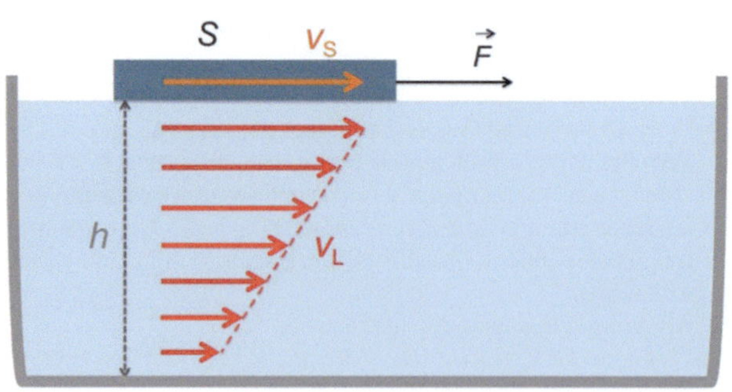

Fig. 8.1 Measuring the dynamic viscosity coefficient of a liquid (see text). S contact surface between the plate and the liquid; F force; h height; v_S plate speed; v_L speed of liquid layers

speed v of the plate itself, while the layers below will slide on each other with a gradually decreasing speed (due to friction), up to the layer near the bottom of the container, which is motionless. Experimentally, it is found that the friction force F_f is directly proportional to the surface S and the speed v of the plate, and inversely proportional to the height h of the liquid in the container:

$$F_f = \mu S \frac{v}{h} \tag{8.1}$$

where μ is the so-called *dynamic viscosity coefficient*, which depends on the fluid and varies (exponentially) with *temperature* (for most liquids, the lower the temperature, the higher the viscosity, while for gases viscosity increases with temperature). As an example, at 0 °C, the viscosity coefficient of mercury is about 1.5 times that of water.

It can be said that viscosity is a measure of how much a fluid resists to flow (or to deformation by *shear stress*). While in liquids it is due, as mentioned, to the sliding of superimposed layers, viscosity of gases is due to the collisions among the gas molecules, and that's why it increases with temperature (according to the kinetic theory of gases discussed in Chap. 2, Box 2.2).

8.1.2 Hagen–Poiseuille Equation

Because of viscosity, a real fluid flows through a tube only if there is a *pressure difference* between its two ends. In fact, it is necessary to do a work (and, hence, apply a net force) against the friction forces in order to maintain the fluid in motion (that's why we need a heart!). Moreover, if the fluid speed is not too high (see Box 8.1), the flow of a real fluid is *laminar* (Fig. 8.2), that is, it occurs in parallel concentric layers which slide on one another with a parabolic velocity profile; the fluid speed is maximum in the middle and decreases gradually in the outer layers (something that is easy to understand considering what we said in Sect. 8.1.1).

Box 8.1: Laminar vs. Turbulent Flow

It can be predicted if the flow of a fluid through a conduit would be laminar or *turbulent* (i.e., with chaotic changes in pressure and speed) using the so-called *Reynolds number* (Re):

$$\mathrm{Re} = \frac{2r\mathrm{V}d}{\mu} = \frac{2r\mathrm{V}}{\nu}$$

where r is the radius of the conduit, V is the mean speed of the fluid, d is fluid density, μ is its dynamic viscosity coefficient, and ν is the so-called *kinematic viscosity coefficient*, i.e., the ratio between dynamic viscosity coefficient and density ($\nu = \mu/d$).

The flow in a cylindrical conduit will be mostly laminar if Re is below a "critical" value of about 2000, while it will be mostly turbulent if it is Re > 4000 (for intermediate values, the two types of flow coexist).

Accordingly, the flow of a given fluid (with its own viscosity and density) in a pipe will become turbulent when its speed will be higher than a certain value. The "sounds" coming from heart, lungs and arteries, which we listen to with our stethoscopes, are always caused by turbulent flow. As an example, where the carotid artery diameter is significantly reduced due to an atherosclerotic plaque, blood velocity increases (according to continuity equation) and the flow becomes turbulent, so that the typical "bruit" can be heard. Note that the "critical" Re value is lower than 2000 in blood vessels, due to complex vascular geometry and pulsatile flow patterns [1].

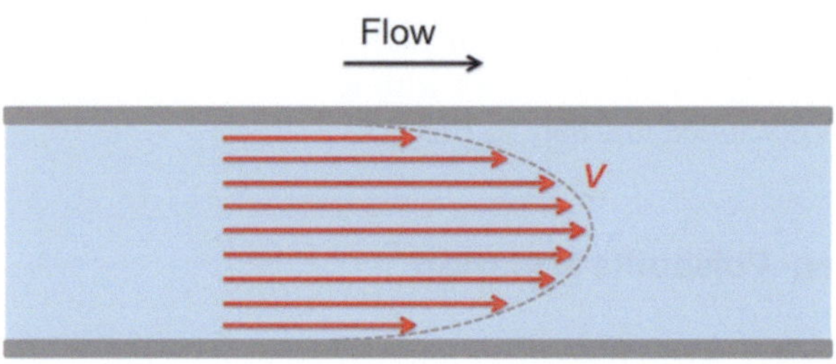

Fig. 8.2 Speed (v) profile of laminar flow. To better understand this figure, look again at Fig. 7.1 (in which the free surface of the liquid has the maximum speed, while the bottom of the container is motionless), and consider that here the whole pipe wall and, accordingly, the upper and lower ends in the two-dimensional representation, are motionless

Hagen–Poiseuille equation (or simply "Poiseuille equation") states that, for a viscous fluid which moves with laminar flow through a cylindrical conduit of radius r and length l (such as the endotracheal tube in Fig. 8.3), the volume flow rate Q (as defined in Chap. 7) is directly proportional to the pressure difference ΔP between the two ends of the conduit and to the fourth power of its radius, and inversely proportional to the viscosity μ of the fluid and to the length of the conduit:

$$Q = \Delta P \frac{\pi r^4}{8\mu l} \tag{8.2}$$

Accustomed as we are, as anesthesiologists or intensivists, to deal with any type of "conduit" (endotracheal tubes, venous catheters, cannulae for Extracorporeal Membrane Oxygenation, and so on), Hagen–Poiseuille equation is probably the physical law to which we are faced more often in our daily clinical practice. Some examples are discussed in the next section. Moreover, as shown in the final part of this chapter, the fundamental relationship between cardiac output and arterial pressure, on which we base hemodynamic monitoring and management, also comes from Hagen–Poiseuille equation.

$$Q = \frac{\Delta P \, \pi \, r^4}{8 \, \mu \, l}$$

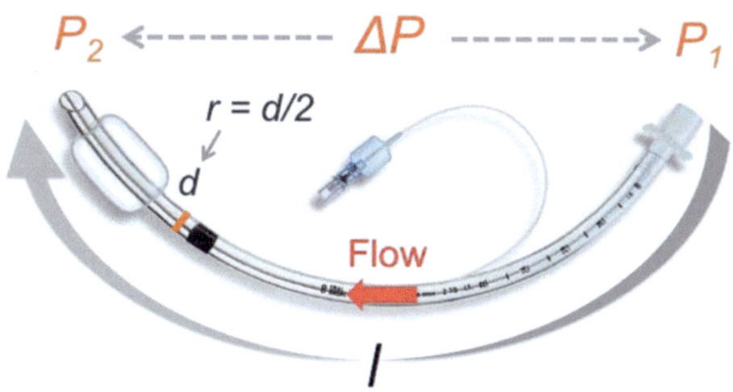

Fig. 8.3 Hagen–Poiseuille equation (for an endotracheal tube). Q volume flow rate; $\Delta P = P_1 - P_2$ pressure difference between the two ends; r radius; d diameter; μ dynamic viscosity coefficient (of air); l length

8.2 Tubes, Cannulae, and Catheters: Some Implications of Hagen–Poiseuille Equation

When we administer intravenous fluids to a patient, the flow rate through the venous catheter is, according to Hagen–Poiseuille equation, directly proportional to the pressure difference between the solution bag and the vein, which is, in turn, proportional to the height of the bag with respect to the patient, according to *Stevin's law* or, more generally, to Bernoulli's theorem (see Chap. 9, Sect. 9.1.2). If rapid fluid resuscitation is required, a shorter intravenous catheter is preferable to a longer one of the same gauge [2, 3]. In fact, according again to Hagen–Poiseuille equation, the volume flow rate produced by the same pressure difference will be higher through a lower length catheter. Hence, although a triple-lumen central venous catheter usually includes a 16-gauge lumen, a 16-gauge peripheral venous catheter will allow a faster fluid administration since it is much shorter than an eight-inch (about 20 cm) central venous catheter. Moreover, we should not be surprised if packed red blood cells, as well as high-concentration glucose solutions, drip more slowly than saline (or another crystalloid solution) through a venous catheter of the same diameter and length; in fact, their viscosity is much higher than that of saline.

It is also clear from Hagen–Poiseuille equation why a so high pressure of your thumb is needed to inject a liquid through an epidural catheter or a pulmonary artery catheter, which are tight and very long.

8.2.1 Endotracheal Tubes, Tracheotomy Cannulae, and Work of Breathing

Hagen–Poiseuille equation also applies to the mixture of oxygen and air which flows through an endotracheal or tracheotomy tube in a mechanically ventilated patient, explaining how the tube dimensions (diameter and length) affects the *work of breathing* (*WOB*) and, accordingly, may favor (or counteract) the weaning from mechanical ventilation.

The definition of WOB simply comes from the rough application of the physical notion of *mechanical work* (see Chap. 6, Box 6.1) to pulmonary ventilation. Imagine the lung as a cylinder of base area A and height x, which expands only in the direction of its height during inspiration (Fig. 8.4). If the height x becomes $x + \Delta x$ at the end of inspiration, then the lung volume will be increased by $\Delta V = A\Delta x$ (since, as mentioned in Chap. 1, the volume of a cylinder is the product of its base area and its height). Called F the force which, acting perpendicularly to A, produces the "displacement" Δx (evidently, along the same direction of F), the work W done by this force, as we have already seen in Chaps. 6 and 7, is:

$$W = F\,\Delta x \tag{8.3}$$

In other words, W is the *energy* "expended" to increase the volume of the lung/cylinder by ΔV. Remembering that pressure is (normal) force per unit area (see Chap. 9), the pressure exerted on A will change by $\Delta P = F/A$. Accordingly, we can replace F with the product $\Delta P\,A$ in Eq. (8.3), which becomes:

$$W = \Delta P\,A\,\Delta x \tag{8.4}$$

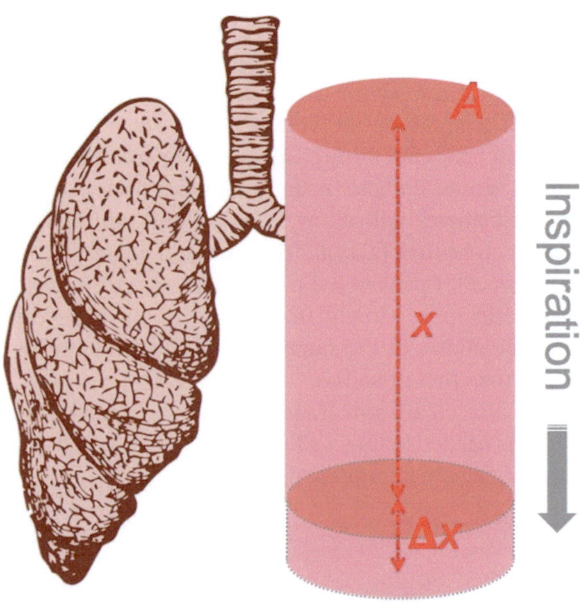

Fig. 8.4 In order to define work of breathing, the lung can be represented as a cylinder of base area A and height x which expands only in the direction of its height (which increases by Δx) during inspiration

Since, as mentioned above, is $\Delta V = A\ \Delta x$ and, hence, is $\Delta x = \Delta V/A$, Eq. (8.4) can be rewritten as:

$$W = \Delta P\, A \frac{\Delta V}{A} = \Delta P\, \Delta V \tag{8.5}$$

which is the definition of (inspiratory) WOB well-known to intensivists and anesthesiologists, roughly "pressure x volume" (there are more rigorous definitions of WOB, but for our purpose we can settle for the simple one provided by Eq. 8.5).

One of the reasons why tracheotomy reduces WOB in mechanically ventilated patients is that tracheotomy cannulae are usually shorter than endotracheal tubes [4]. In fact, as it results from Eq. (8.5), WOB is proportional to the pressure difference ΔP which is needed to develop the tidal volume (the so-called *driving pressure*) and, according to Hagen–Poiseuille equation, the patient will need a lower ΔP to receive the same gas flow if the length of the tube is lower and, thus, he/she will do less work.

Furthermore, a larger internal diameter tracheotomy or endotracheal tube (ETT) may facilitate (and accelerate) the weaning from mechanical ventilation. In fact, for a given pressure difference, the patient will receive a much greater gas flow through a slightly larger tracheotomy cannula or ETT (since flow depends on the fourth power of radius) and, accordingly, the same flow will be obtained with a much lower driving pressure, again reducing WOB [5].

8.2.2 Cannulae for Extracorporeal Membrane Oxygenation: Hagen–Poiseuille Equation and the Pseudoplastic Behavior of Blood

Of course, the diameter and the length of both venous (drainage) and arterial (return) cannulae are key determinants of flow in Extracorporeal Membrane Oxygenation (ECMO), according to Hagen–Poiseuille equation [6, 7]. For example, with the same flow, the "pressure drop" will be much lower in a larger diameter cannula. Moreover, the dimensions of the cannula (particularly, of the arterial cannula) affect the flow limit beyond which flow changes from laminar to turbulent (see Box 8.1) and, accordingly, Hagen–Poiseuille equation does no longer apply (since, as stated above, it is only valid for a fluid which moves with laminar flow) and blood trauma increases [6]. However, it was recently shown that pressure-flow charts provided by most ECMO cannulae manufacturers (usually using water as testing fluid) do not accurately describe what occurs, with blood, in clinical practice [6, 7]. This is mainly because blood, due to the presence of red blood cells, behaves as a *non-Newtonian fluid*, particularly a *pseudoplastic* fluid. Unlike *Newtonian fluids* such as water, which have a characteristic viscosity (at a given temperature), the viscosity of non-Newtonian fluids depends on the "intensity" of the shear stress which is applied on them (expressed as *shear rate*, i.e., the rate at which a shearing deformation is applied to a body, or the speed difference between two points of the body divided by the distance between them) [8]. In particular, viscosity increases with stress for the so-called *shear thickening* (or *dilatant*) non-Newtonian fluids (such as the suspensions of corn starch in water which allow people to "walk on water," as sometimes seen on TV; in fact, these suspensions behave as liquids if crossed slowly, but become very viscous when trampled on strongly while running [9]); conversely, viscosity decreases with stress for *shear thinning* (or *pseudoplastic*) non-Newtonian fluids such as toothpaste, ketchup, mayonnaise, ultrasound gel (to give a "clinical" example) and, as said, blood. As experience shows, it is easier to pour ketchup on a hamburger, or ultrasound gel on the probe (see Chap. 20), if you shake the bottle vigorously before; in fact, doing so, you apply a high shear rate to the (pseudoplastic) fluid and, accordingly, its viscosity decreases (so that it comes out from the tight mouth of the bottle more easily) [10]. Similarly, the viscosity of blood decreases with its speed, and this explains the disagreement between the pressure-flow charts provided by ECMO cannulae manufacturers and the flows (or the pressure drops for the same flows) observed when human blood, instead of water, is used to test the cannulae [6, 7]. Moreover, although the velocities at which blood flow becomes turbulent are usually higher as compared with water since blood viscosity is higher than that of water (just have a look to the Reynolds number formula in Box 8.1 to realize this), the reduction in blood viscosity at very high velocities (e.g., due to a too narrow cannula) may lead to turbulent flow at different speeds that those predicted by tests with water [6].

8.3 Hagen–Poiseuille Equation and Hemodynamics

If we call *resistance* (R) the quantity $8\ \mu l/\pi r^4$ (flow, in fact, is hindered by the increase in length and viscosity, and favored by the increase in radius), Eq. (8.2) becomes:

$$Q = \frac{\Delta P}{R} \tag{8.6}$$

or

$$\Delta P = Q\,R \tag{8.7}$$

In the latter form, Hagen–Poiseuille equation is sometimes referred to as the *Ohm's law of hydrodynamics*, due to its analogy to *Ohm's law for electric current*:

$$P.D. = i\,R_e \tag{8.8}$$

where *P.D.* is the electric potential difference (i.e., the voltage, a concept similar to the pressure difference), i the electric current intensity (defined similarly to the flow rate), and R_e the electrical resistance (see Chap. 11) [1, 11]. When applied to the cardiovascular system, these equations express the well-known relationship between arterial pressure and cardiac output (CO) and define systemic (or pulmonary) vascular resistance:

$$\text{MAP} - \text{CVP} = \text{CO} \times \text{SVR} \text{ or } \text{SVR} = \frac{\text{MAP} - \text{PVC}}{\text{CO}} \tag{8.9}$$

$$\text{MAP} - \text{PAOP} = \text{CO} \times \text{PVR} \text{ or } \text{PVR} = \frac{\text{MPAP} - \text{POAP}}{\text{CO}} \tag{8.10}$$

where MAP and MPAP are, respectively, mean arterial pressure and mean pulmonary artery pressure, CVP is central venous pressure, PAOP is pulmonary artery occlusion pressure, and SVR and PVR are, respectively, systemic and pulmonary vascular resistances. Cardiac output monitors use Eqs. (8.9) and (8.10) to display a "calculated value" of SVR and PVR, respectively.

We usually identify vascular resistance with the vascular "tone," namely the degree of vasodilation/vasoconstriction (which is, of course, its main determinant to be considered in clinical practice). However, this is a simplified way to define vascular resistance. In particular, it is clear from Eq. (8.2) that vascular resistance may also vary with blood viscosity which, as mentioned in the previous section, is not constant [12]. Indeed, in addition to increase at lower flow rates due to the non-Newtonian (pseudoplastic) behavior of blood (see Sect. 8.2.2), what that affects especially microcirculation [1, 8], blood viscosity is inversely related to *temperature* (as for liquids in general, as discussed in Sect. 8.1.1) and also increases with *hematocrit* [1]; both these effects can become significant during (hypothermic) cardiopulmonary bypass.

References

1. Crystal GJ, Heerdt PM. Cardiovascular physiology: integrative function. In: Hemmings Jr HC, Egan TD, editors. Pharmacology and physiology for anesthesia: foundations and clinical application. Elsevier Saunders; 2013. p. 366–89.
2. Reddick AD, Ronald J, Morrison WG. Intravenous fluid resuscitation: was Poiseuille right? Emerg Med J. 2011;28:201–2.
3. Jayanthi NV, Dabke HV. The effect of IV cannula length on the rate of infusion. Injury. 2006;37:41–5.
4. Diehl JL, El Atrous S, Touchard D, Lemaire F, Brochard L. Changes in the work of breathing induced by tracheotomy in ventilator-dependent patients. Am J Respir Crit Care Med. 1999;159(2):383–8.
5. O'Callaghan DJ, Wyncoll D. What size tube doctor? Bigger may be better – at least for weaning. Crit Care. 2013;17:422.
6. Broman LM, Prahl Wittberg L, Westlund CJ, et al. Pressure and flow properties of cannulae for extracorporeal membrane oxygenation I: return (arterial) cannulae. Perfusion. 2019;34(1S):58–64.
7. Broman LM, Prahl Wittberg L, Westlund CJ, et al. Pressure and flow properties of cannulae for extracorporeal membrane oxygenation II: drainage (venous) cannulae. Perfusion. 2019;34(1S):65–7.
8. Pop GA, Duncker DJ, Gardien M, Vranckx P, Versluis S, Hasan D, Slager CJ. The clinical significance of whole blood viscosity in (cardio)vascular medicine. Neth Heart J. 2002;10(12):512–6.

9. Crawford NC, Popp LB, Johns KE, et al. Shear thickening of corn starch suspensions: does concentration matter? J Colloid Interface Sci. 2013;396:83–9.
10. Walker J. The flying circus of physics. 2nd ed. Wiley; 2007. p. 105–6.
11. Xu D, Olivier NB, Mukkamala R. Continuous cardiac output and left atrial pressure monitoring by long time interval analysis of the pulmonary artery pressure waveform: proof of concept in dogs. J Appl Physiol. 2009;106:651–61.
12. McGregor M, Sniderman A. On pulmonary vascular resistance: the need for more precise definition. Am J Cardiol. 1985;55:217–21.

Toothpaste, Sea Deeps, and Invasive Pressure Monitoring: Stevin's Law and Pascal's Principle

9

Contents

Where's Physics

Daily life	"Under the sea" (or "Up in the sky") Squeezing toothpaste Lift a car (in a mechanical workshop) Save the life of someone who is choking (or who is in cardiac arrest)
Physics involved	Hydrostatic pressure Stevin's law Pascal's principle Variation in atmospheric pressure with altitude
Clinical practice	Small and large syringes Verify the inflation of a tracheal tube cuff Heimlich maneuver External cardiac massage Invasive pressure monitoring Measuring pulmonary artery occlusion pressure Height of the pressure transducer relative to the patient (leveling) Pressure transducer zeroing Phleboclysis

A. Pisano, *Physics for Anesthesiologists and Intensivists*,
https://doi.org/10.1007/978-3-030-72047-6_9

9.1 Fluids at Rest: Stevin's Law and Pascal's Principle

Chapters 7 and 8 have dealt with the motion of fluids, i.e., that part of *fluid mechanics* which is known as *fluid dynamics*. The so-called *statics of fluids*, on the contrary, deals with fluids which are "at rest." Below we will discuss two simple laws of fluid statics which are the basis of invasive pressure monitoring.

First, though, it is useful to recall the definitions of, and to say something more about, two physical quantities which we have encountered several times in previous chapters: *density* and *pressure*.

9.1.1 Density and Pressure of Fluids

The *density* d of a fluid sample of mass m and volume V, *which has a uniform density*, is:

$$d = \frac{m}{V} \tag{9.1}$$

or the mass per unit volume. Of course, according to what discussed in Chap. 2, while the density of a liquid is (substantially) constant, that of a gas depends largely on pressure and temperature (in fact, the volume of a given mass of any gas varies with these two properties, according to the ideal gas law).

Practically, the density of a substance is often reported as its *specific gravity*, i.e., the ratio between its density and that of water (which is around 1000 kg/m^3).

In general, a force F which acts perpendicularly (and *uniformly*) on a flat surface A exerts a pressure p which can be defined as:

$$p = \frac{F}{A} \tag{9.2}$$

that is the force per unit area (accordingly, the pressure unit in the International System of Units is Newton per square meter, i.e., the unit of force divided by the unit of area, although many other pressure units are used in practice, as discussed in Box 9.1).

Box 9.1: Pressure Units and Jokes

There are many different pressure units, which reflect the research context in which they were developed (as the most convenient in that setting). In the International System of Units (SI), pressure is measured in *Pascal* (Pa), i.e., the pressure that a force of 1 Newton (the SI unit of force) exerts on a 1 square meter surface (N/m^2). A funny reminder of that is provided by a joke which circulates among scientists and students in the physics or engineering setting, and that really can only make them and some other "quasi-geek" like me laugh. It goes something like this (you can find slightly different versions on the internet):

Newton, Pascal, and Archimedes [at other times the third physicist is Einstein] *are playing a game of hide and seek. While Archimedes counts to ten, Pascal runs away and hides. Conversely, Newton draws with chalk a one meter by one meter square on the ground, and sits inside it. When Archimedes opens his eyes, he says: "Newton... I found you!", but Newton answers: "I'm not Newton... I'm Pascal!"* (if you got it, you will never forget that 1 Pascal is 1 Newton per square meter...).

Jokes aside, the Pascal is a very small unit, which is rather inconvenient in most contexts. In medicine, we mostly use *millimeters of mercury* (mmHg or *Torr*), i.e., the pressure exerted by a 1 mm high column of mercury, although kilopascal (1 kPa = 1000 Pa) is also used in some countries, e.g., as unit of partial pressure of blood gases (1 mmHg $\approx$ 0.133 kPa). For compressed gases, e.g., oxygen in cylinders for medical use, the pressure unit is usually the *bar* (1 bar $\approx$ 1 atmosphere [atm] = 760 mmHg = 101, 3 kPa, i.e., the atmospheric pressure at sea level). Other pressure units include pounds per square inch (psi) and millibar (1 mbar = 100 Pa $\approx$ 0.75 mmHg), mostly used in the industry and in meteorology, respectively.

Pressure within a fluid can be due to the weight of the fluid itself (i.e., to *gravity*), acceleration (we saw in Chap. 7 how the pressure of a fluid in motion changes as its speed varies, according to Bernoulli's theorem), and external forces applied to the fluid. In particular, the pressure which is exerted by a fluid *at rest* due to its weight (which, as mentioned in Box 3.1, is the *force* which acts on a body due to gravity) is called *hydrostatic pressure* and, as we will see shortly, depends on the depth within the fluid and on fluid density (the latter is rather intuitive since, as discussed in Chaps. 3 and 14, weight is proportional to mass and, as just mentioned, density is none other than "the amount of mass" contained in a given volume).

The well-known notion that smaller syringes develop higher pressure as compared with larger ones with the same force applied to the plunger (here is a case of pressure generated by an external force) simply comes from the definition of pressure provided by Eq. (9.2); in fact, the same force acts on a smaller surface area in a smaller syringe, as shown in Fig. 9.1. This explains, for example, the common recommendation to use at least 10 mL syringes to flush long-term central venous catheters since smaller syringes may increase the risk of catheter rupture (due to the higher pressure developed) if resistance is encountered [1].

Note that in the definitions of density and pressure (Eqs. 9.1 and 9.2) we assumed, respectively, that the force is "uniformly applied" and, similarly, the density is "uniform." Otherwise, we should resort to the mathematical concept of *limit* (see Chap. 1) for a correct (and more general) definition of these quantities. For example, the density of a fluid is, more rigorously, the limit of the ratio between its mass and

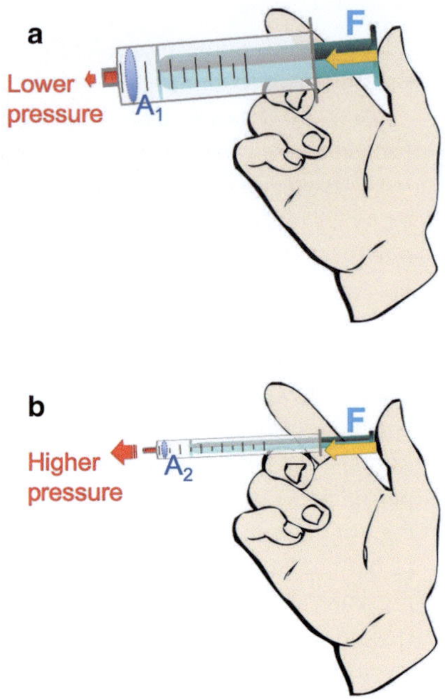

Fig. 9.1 According to the definition of pressure as force per unit area, the same force *F* applied with your thumb to the plunger of a larger syringe (**a**) results in a lower pressure exerted on the fluid contained therein as compared with that generated in a smaller syringe (**b**). In fact, the surface area A_1 on which the force acts in the higher volume syringe is larger than the surface area A_2 on which it acts in the smaller syringe

volume for an infinitesimal volume of that fluid (however, "mass divided by volume" is enough for our purposes!).

9.1.2 Under the Sea: Stevin's Law

Consider a given volume of liquid in a cylindrical container, e.g., a glass of water (Fig. 9.2). The force that this liquid exerts on the container base (which has an area *A*) is nothing but its weight *W*, that is, according to Newton's second law (see Chap. 3, Box 3.1, and Chap. 14):

$$W = mg \tag{9.3}$$

where *m* is the mass of fluid which is contained in the glass and *g* is free-fall acceleration.

The hydrostatic pressure p_i which is exerted on the base area *A* is, hence (Eq. 9.2):

$$p_i = \frac{W}{A} = \frac{mg}{A} \tag{9.4}$$

According to Eq. (9.1), mass is the product of volume and density. Moreover, as mentioned in Chap. 1, the volume *V* of a cylinder with base area *A* and height *h* is the product of *A* and *h*. After these substitutions, Eq. (9.4) becomes:

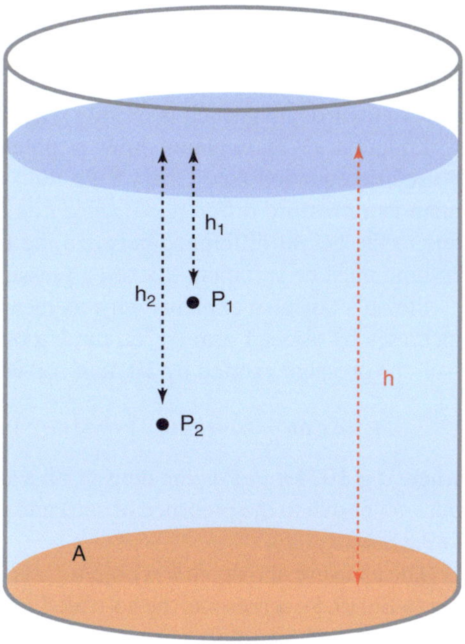

Fig. 9.2 The hydrostatic pressure that a certain volume of a given liquid exerts on the base area *A* of its container depends only on the height *h* of the liquid "column" (Stevin's law). Accordingly, the pressure P_1 at any point which lies at a depth h_1 is lower than the pressure P_2 at any point which is located at a greater depth h_2; in fact, a fluid column of greater height "pushes" above the latter

$$p_i = \frac{dVg}{A} = \frac{dAhg}{A} = dgh \tag{9.5}$$

where *d* is the density of the liquid.

Equation (9.5) is known as *Stevin's law*. It states that the pressure which a column of a given fluid exerts on its base depends on its height but not on any horizontal dimension (and, hence, on the shape of the container). It goes without saying that, accordingly, the pressure at any point within a fluid at rest depends only on the depth *h* of that point (see Fig. 9.2).

It is interesting to note that Stevin's law is no more than a particular case of Bernoulli's theorem, that is the physical law which, at the same time, explains how an airplane can fly and how the Venturi mask works, allows echocardiographic measurement of pressure gradients, and is involved in the pathophysiology of dynamic left ventricular outflow tract obstruction due to systolic anterior motion (SAM), as discussed in Chap. 7. Considering again Fig. 9.2, Eq. (7.12) (Chap. 7) for two points at height h_1 and h_2 (where pressures are, respectively, p_1 and p_2) can be written as follows:

$$d\,g\,h_1 + p_1 = d\,g\,h_2 + p_2 \tag{9.6}$$

In fact, since the liquid is at rest, its speed is zero and, accordingly, the quantities $\frac{1}{2}dv_1^2$ and $\frac{1}{2}dv_2^2$ are both equal to zero.

After a simple rearrangement, Eq. (9.6) becomes:

$$p_1 - p_2 = d\,g\,h_2 - d\,g\,h_1 = d\,g\left(h_2 - h_1\right) = d\,g\,\Delta h \tag{9.7}$$

that is, the pressure difference between the two points is proportional to the difference in their depth (which is exactly what Stevin's law states).

Equation (9.7) explains how a phleboclysis works. According to Hagen–Poiseuille equation (see Chap. 8), a fluid flows through a venous catheter only if there is a pressure difference between its two ends. Such a difference is achieved due to the height difference between the bag and the vein (the height of the fluid column must be sufficient to exert a pressure higher than the venous pressure).

Stevin's law also explains why, as mentioned in Chap. 5, undersea the pressure increases by about 1 atm (≈760 mmHg or 101.3 kPa) for every 10 m of depth. In fact, the pressure exerted by 10 m of sea water is:

$$p = d\,g\,\Delta h = 1024\left(\mathrm{kg\,m^{-3}}\right)\times 9.8\left(\mathrm{ms^{-2}}\right)\times 10\,\mathrm{m} \approx 100{,}000\,\mathrm{Pa} = 100\,\mathrm{kPa} \approx 1\,\mathrm{atm} \tag{9.8}$$

where d = 1024 kg/m^3 is the density of sea water (which is a bit higher than that of pure water due to the presence of salt) and g, as stated in Chap. 3 (Box 3.1), is equal to 9.8 m/s^2.

The pressure at a depth h within a glass of water, or undersea, calculated according to Eq. (9.5), represents the so-called *gauge pressure*, i.e., the difference between the "absolute" (or total) pressure and atmospheric pressure. In fact, the pressure exerted by air (i.e., the atmospheric pressure p_{Atm}) must be added to that exerted by water. Accordingly, the *absolute pressure* (p_{tot}) at any depth h within a liquid is:

$$p_{\mathrm{tot}} = p_{\mathrm{Atm}} + d\,g\,h \tag{9.9}$$

As mentioned in Chap. 2, atmospheric pressure (which is 760 mmHg or 101.3 kPa at sea level) decreases as altitude increases; this, however, occurs only in part according to Stevin's law (see below).

9.1.3 Push, Squeeze, and Lift: Pascal's Principle

A variation in the pressure applied to an enclosed fluid is transmitted unchanged to each portion of the fluid and to the walls of its container, where it acts perpendicularly. This is the *Pascal's principle,* so called after Blaise Pascal, from whom also derives the name of the pressure unit of the International System of Units (SI), i.e., Newton/m^2 or pascal (Pa) (see Box 9.1).

Thanks to Pascal's principle, it is possible to push out toothpaste by squeezing the tube at its bottom. Another useful application of this principle can be found in a mechanical workshop; in a hydraulic car lift (or, in general, *hydraulic lever*), for example, the Pascal's principle is used to magnify a force so much to allow lifting a car. Figure 9.3 shows a schematic representation of such a machinery. According to Pascal's principle, a pressure applied to the surface area A_0 on the right is transmitted unchanged to the surface area A_1 on the left. Remembering the definition of pressure (Eq. 9.2), it is therefore:

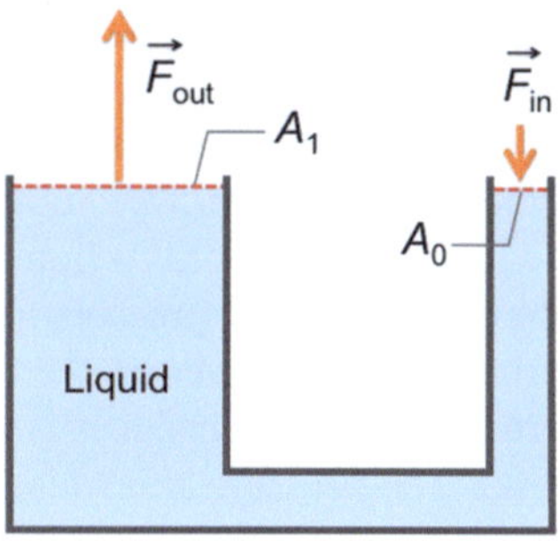

Fig. 9.3 Scheme of a hydraulic lever (see text). F_{in} Input force, which is applied to the surface area A_0; F_{out} Output force (exerted by the surface A_1)

$$\frac{F_{in}}{A_0} = \frac{F_{out}}{A_1} \tag{9.10}$$

Since A_0 is smaller than A_1, the output force F_{out} will be higher than the input force F_{in}. The "trick" is that the input force must be applied for a greater distance than that along which the output force applies. Accordingly, the *work* done by the two forces is the same (see Chap. 6, Box 6.1). However, the weight which we can lift is greater than that which we could lift with our own "force" (and this is what matters).

Furthermore, the Pascal's principle may save a life. Heimlich maneuver relies precisely on it; a pressure properly applied to the abdomen is transmitted undiminished to the thorax and then to the trachea, pushing out the foreign body [2]. Similarly, the abdomen gets up rhythmically during *external cardiac massage* (thanks to which the blood is pushed into the circulation, again according to Pascal's principle).

For a fluid whose weight is negligible, e.g., the air contained in an empty bottle (which, evidently, is not at all "empty" but full of air!), Pascal's principle tells us that *the pressure is anywhere the same*. In fact, this law can be also enunciated as follows; in the absence of *volume forces* (i.e., forces which act on all particles of a body, and are accordingly proportional to its volume: a typical example is exactly the weight), the pressure is constant at any point within the fluid.

Since the density of air (at 20 °C and 1 atm) is about 1.2 kg/m^3 (i.e., more than 800 times lower than that of water), we can certainly neglect the weight of the small amount of air that we inject into the cuff of an endotracheal tube (or of a supraglottic airway device). Accordingly, the pressure in the cuff is the same of that in the pilot balloon. Moreover, the pressure exerted by the tracheal wall on the cuff, which according to *Newton's third law* is equal and opposite to that exerted by the cuff on the tracheal wall (see Chaps. 2 and 14), is transmitted unchanged to the pilot balloon [3]. Hence, we can check the correct inflation of the cuff through the pilot balloon (we have already seen in Chap. 6 that it would be better to do it with a pressure gauge rather than "manually") [4–9].

Finally, it is again Pascal's principle which makes invasive pressure monitoring possible, as discussed below.

9.2 Invasive Pressure Monitoring

When we insert a catheter into an artery (e.g., the radial artery), the "system" comprising the catheter, the tubing, and the *pressure transducer* behaves as a closed container which holds a liquid (usually normal saline). Accordingly, any change in the arterial blood pressure is transmitted unaltered into the transducer (Pascal's principle). Moreover, if the artery and the transducer are at the same height, the pressure in the transducer is just the same that in the artery (according to Stevin's law). The pressure transducer consists of a membrane through which an electric current runs; the pressure and its changes within the transducer distort the membrane, leading to changes in its *electrical resistance* (see Chap. 11) [10]. These (very small) changes in electrical resistance can be measured with high precision by means of a special electric circuit (the so-called *Wheatstone bridge*, which is discussed in detail in Chap. 11) [11, 12], and are then displayed as a waveform which accurately reproduces the "original" arterial pressure waveform through a mathematical method known as *Fourier analysis*, according to which any complex waveform can be represented as the sum of simpler sine waves (called *harmonics*) whose frequencies are multiples of the fundamental frequency (all will be clearer in Chap. 19, which deals with oscillations, mechanical waves, and resonance and where, moreover, you will find other important considerations about the systems for invasive pressure monitoring) [11–14].

Of course, all this also applies to other pressures such as *central venous pressure* (CVP) or pulmonary artery pressure. Moreover, thanks again to Pascal's principle, we can measure left atrial pressure with a catheter whose tip is very far from the left atrium. In fact, after occluding a pulmonary artery branch by inflating the balloon of a pulmonary artery catheter, the pressure that we measure downstream, i.e., the *pulmonary artery occlusion pressure* (PAOP), is (approximately) the same that in the left atrium, since there are no valves in the pulmonary vasculature between pulmonary artery and left atrium.

9.2.1 Leveling: How Important Is the Difference?

Everyone of us knows very well that the correct *leveling* of the transducer (usually at the mid-thoracic position) is pivotal when measuring pressures with a fluid-filled system, such as invasive arterial pressure or CVP [13, 15, 16]. As an example, if the transducer lies ten centimeters below the point at which the measurement should be made, according to Stevin's law the (hydrostatic) pressure exerted by a 10 cm high "column" of normal saline is added to the patient's true pressure, which will be accordingly overestimated. Conversely, the pressure will be underestimated if the transducer is located too high. In particular, every centimeter of height difference leads to an error of 0.74 mmHg: that is, if the transducer lies 10 cm above the correct level, the pressure will be underestimated by 7.4 mmHg. This is due to the different densities of mercury and normal saline [10, 16], as it can be easily shown by applying, again, Stevin's law.

According to Eq. (9.5), in fact, we can write:

$$d_w \, h_w \, g = d_{Hg} \, h_{Hg} \, g \tag{9.11}$$

or

$$\frac{h_w}{h_{Hg}} = \frac{d_{Hg}}{d_w} \tag{9.12}$$

where d_w and d_{Hg} are, respectively, the density of water (which approximates that of normal saline) and the density of mercury, and h_w is the height of a column of water that exerts the same hydrostatic pressure which is exerted by a column of mercury of height h_{Hg}. Hence, the conversion factor between cmH_2O and cmHg (beware, not mmHg) is equal to the ratio between the density of mercury (which is about 13,600 kg/m^3) and that of water, i.e., 13,600/1000 = 13.6 (that is the *specific gravity* of mercury). Finally, to convert cmHg to mmHg you have to multiply h_{Hg} by 10, then we have 1 mmHg = 13.6/10 = 1.36 cmH_2O, and 1 cmH_2O = 1/1.36 ≈ 0.74 mmHg. Which is what we said.

It is clear that a correct transducer leveling is particularly important when measuring filling pressures (CVP and PAOP). In fact, an error of 5–10 mmHg (due to a height difference of 7–14 cm, that are not so many!) is usually of little clinical relevance when measuring arterial blood pressure. Conversely, it may lead to big "blunders" when dealing, for example, with CVP, whose values (and changes) are most often in that order [13, 15, 16].

9.2.2 Differences that Matter and Really Insignificant Differences: Zeroing

All pressures to which we usually refer in clinical practice are, actually, *gauge pressures* (see Sect. 9.1.2). When you say: "Mr. Smith, your systolic pressure is 140 mmHg," the absolute systolic pressure in Mr. Smith's artery is, indeed, 900 mmHg (at sea level), i.e., 140 mmHg higher than the atmospheric pressure (760 mmHg or 101.3 kPa) [10, 13]. Zeroing the transducer means "tell" the monitoring system how much is the atmospheric pressure, so that it can subtract that value to the measured pressure in order to display the gauge pressure (i.e., the extra pressure as compared to the atmospheric pressure). Invasive pressure monitoring requires "manual" zeroing since the manufacturer cannot know at which atmospheric pressure the monitoring device will be used, and there can be not negligible differences in atmospheric pressure (i.e., the "zero") even within relatively small distances. For example, the hospital where I work is 180 m above sea level (in a big seaside town): at this altitude, atmospheric pressure is about 743 mmHg, i.e., 17 mmHg lower than at sea level; accordingly, if a patient is transferred to a nearby hospital at sea level with a transport monitor displaying invasive arterial blood pressure, his/her pressure will result 17 mmHg higher (perhaps giving the false impression that the patient is getting better) if you don't repeat zeroing! Of course, the error will be removed when the patient will be connected to the new hospital's monitor.

However, does it matter where the transducer is located (relative to the patient) when zeroing it? Absolutely not. Also this can be easily verified using Stevin's law.

In reality, atmospheric pressure does not vary with height (i.e., with altitude) according to Eq. (9.7), but according to an exponential equation which is known as *law of atmospheres*:

$$p = p_0\, e^{-Ah} \tag{9.13}$$

where p is the atmospheric pressure at an altitude h, p_0 is the atmospheric pressure for $h = 0$, e is the base of natural logarithm, and A is an appropriate constant (which includes *density*).

This is due to the decrease in air density with pressure (see above) and, hence, with altitude. However, for small altitude changes (so that the density variation can be neglected), Stevin's law also applies to atmospheric pressure. Of course, we can consider that air density variation is negligible within the same room. Therefore, we can calculate the change in atmospheric pressure Δp_{Atm} for each meter of height change according to Eq. (9.7):

$$\begin{aligned}\Delta p_{\mathrm{Atm}} &= d\, g\, \Delta h = 1.2\left(\mathrm{kg\,m^{-3}}\right) \times 9.8\left(\mathrm{m\,s^{-2}}\right) \times 1\,\mathrm{m} \\ &\approx 11.8\,\mathrm{Pa} = 0.012\,\mathrm{kPa} \approx 0.09\,\mathrm{mmHg}\end{aligned} \tag{9.14}$$

where d = 1.2 kg/m^3 is air density at 20 °C and sea level. That is, an error of 0.0009 mmHg for every variation by 1 cm of the transducer height; it really matters very little were the transducer was during zeroing!

If you are really finicky, though, you'd better redo zeroing when, from the Intensive Care Unit on the ground floor, you bring the patient to Radiology which is, for example, at the fourth floor (a Δh of 12 m translates into a pressure difference of about 1 mmHg).

References

1. Goossens GA. Flushing and locking of venous catheters: available evidence and evidence deficit. Nurs Res Pract. 2015;2015:985686.
2. Srivastava A, Sood A, Joy PS, Mandal S, Panwar R, Ravichandran S, Sarangi S, Woodcock J. Principles of physics in surgery: the laws of mechanics and vectors physics for surgeons-part 2. Indian J Surg. 2010;72:355–61.
3. Hwang JY, Han SH, Park SH, Park SJ, Park S, Oh SH, Kim JH. Interrupting gel layer between Double cuffs prevents fluid leakage past tracheal tube cuffs. Br J Anaesth. 2013;111:496–504.
4. Heffner JE, Hess D. Tracheostomy management in the chronically ventilated patient. Clin Chest Med. 2001;22(1):55–69.
5. Goyal R, Kumar G, Waghray MR. Endotracheal tube cuff pressure monitoring in peripheral hospitals. Med J Armed Forces India. 2006;62(3):243–5.
6. Liu J, Zhang X, Gong W, Li S, Wang F, Fu S, Zhang M, Hang Y. Correlations between controlled endotracheal tube cuff pressure and postprocedural complications: a multicenter study. Anesth Analg. 2010;111(5):1133–7.
7. Michlig SA. Anaesthetic staff cannot identify extremely high tracheal tube cuff pressures by palpation of the pilot balloon. Br J Anaesth. 2013;111(2):300–1.

8. Pisano A. Pitfalls from physics: why we can't "feel" the tube cuff pressure with our fingers. Anesth Analg. 2017;124(4):1368.
9. Pisano A, Verniero L, Galdieri N, Corcione A. Assessing the correct inflation of the endotracheal tube cuff: a larger pilot balloon increases the sensitivity of the 'finger-pressure' technique, but it remains poorly reliable in clinical practice. J Clin Monit Comput. 2019;33(2):301–5.
10. Magder S. Invasive intravascular hemodynamic monitoring: technical issues. Crit Care Clin. 2007;23:401–14.
11. Szocik JF, Barker SJ, Tremper KK. Fundamental principles of monitoring instrumentation. In: Miller RD, editor. Miller's anesthesia. 6th ed. Elsevier Churchill Livingstone; 2005. p. 1191–225.
12. Aston D, Rivers A, Dharmadasa A. Equipment in anaesthesia and critical care. Scion Publishing Limited; 2014. p. 375.
13. Mark JB, Slaughter TF. Cardiovascular monitoring. In: Miller RD, editor. Miller's anesthesia. 6th ed. Elsevier Churchill Livingstone; 2005. p. 1265–362.
14. Pittman JA, Ping JS, Mark JB. Arterial and central venous pressure monitoring. Int Anesthesiol Clin. 2004;42(1):13–30.
15. Magder S. Central venous pressure monitoring. Curr Opin Crit Care. 2006;12:219–27.
16. Magder S. Central venous pressure: a useful but not so simple measurement. Crit Care Med. 2006;34:2224–7.

Part IV

Heat, Temperature, and Electricity: Hemodynamic Monitoring and Much More

Heat, Cardiac Output, and What Is the Future: The Laws of Thermodynamics 10

Contents

Where's Physics

Daily life	Past and future Temperature and thermometers Warming your hands Pans (and handles) Car radiators Fuel = motion + heat (and noise) Fridges and power plants Thermal excursion in the desert
Physics involved	Laws of thermodynamics Thermal capacity and specific heat
Clinical practice	Measuring cardiac output by thermodilution (derivation of the Stewart-Hamilton equation) Core and peripheral temperature Zero-heat-flux thermometer Heat exchangers Vaporizers (see Chap. 15)

A. Pisano, *Physics for Anesthesiologists and Intensivists*,
https://doi.org/10.1007/978-3-030-72047-6_10

10.1 Temperature, Heat, and Energy: The Laws of Thermodynamics

Figure 10.1 shows a number of common objects and situations: (a) what is temperature and how it can be measured; (b) the conversion of the fuel chemical energy into kinetic energy (i.e., movement) and heat in a car engine; (c) warming your hands while keeping a cup of tea; (d) what happens in a fridge is impossible unless you also consider what happens, at the same time, in a power plant; (e) the direction in which time flows (i.e., what distinguishes past from future); (f) the choice of the materials with which a pot is constructed; (g) coolers (e.g., car radiators) and heaters, including heat exchanger of heart-lung machine; (h) measuring cardiac output by thermodilution (e.g., through a pulmonary artery catheter). All this, in a way or another, has something to do with *thermodynamics* and its laws. Hence, the laws of thermodynamics are maybe the best example of physical principles that we face continuously, "from daily life to clinical practice" (as the subtitle of this book). Unfortunately, thermodynamics is a quite complex matter, which is not the case to discuss in details here. Accordingly, below I've only tried to provide some hints about the meaning of the three laws of thermodynamics, as well as of the concept of specific heat capacity, which also are the principles on which the measurement of cardiac output by thermodilution relies. A thorough and systematic discussion on this topic, which is at the same time (wonderfully) easily understandable for non-physicists, can be found in the book *Four laws that drive the Universe* by Peter Atkins [1].

10.1.1 Temperature and Thermometers: The Zeroth Law of Thermodynamics

If a body X is put in close contact with another body Y, and no *heat* is transferred from X to Y or vice versa, it is said that the two bodies are in *thermal equilibrium* with one other. Heat (symbolized Q) is the *thermal energy* of a body, i.e., the "internal energy" due to the random motion of atoms and molecules within it.

If the body Y is, in turn, in thermal equilibrium with another body Z, then also X and Z are in thermal equilibrium with each other. This according to the so-called *zeroth law of thermodynamics*, which states that if two bodies are each in thermal equilibrium with some third body, then they are also in thermal equilibrium with each other. Evidently, these bodies must share a property, that we call *temperature*. In other words, *two objects that are in thermal equilibrium have the same temperature*.

Now we can then define the heat Q as the energy which is transferred from one body to another (or between a "system" and its environment) due to a temperature difference between them. Simply, a cup of hot tea left in the kitchen will cool as some heat is transferred from it to the surrounding environment (whose temperature is lower). The energy transfer between the cup of tea and the environment will continue until both are at the same (room) temperature.

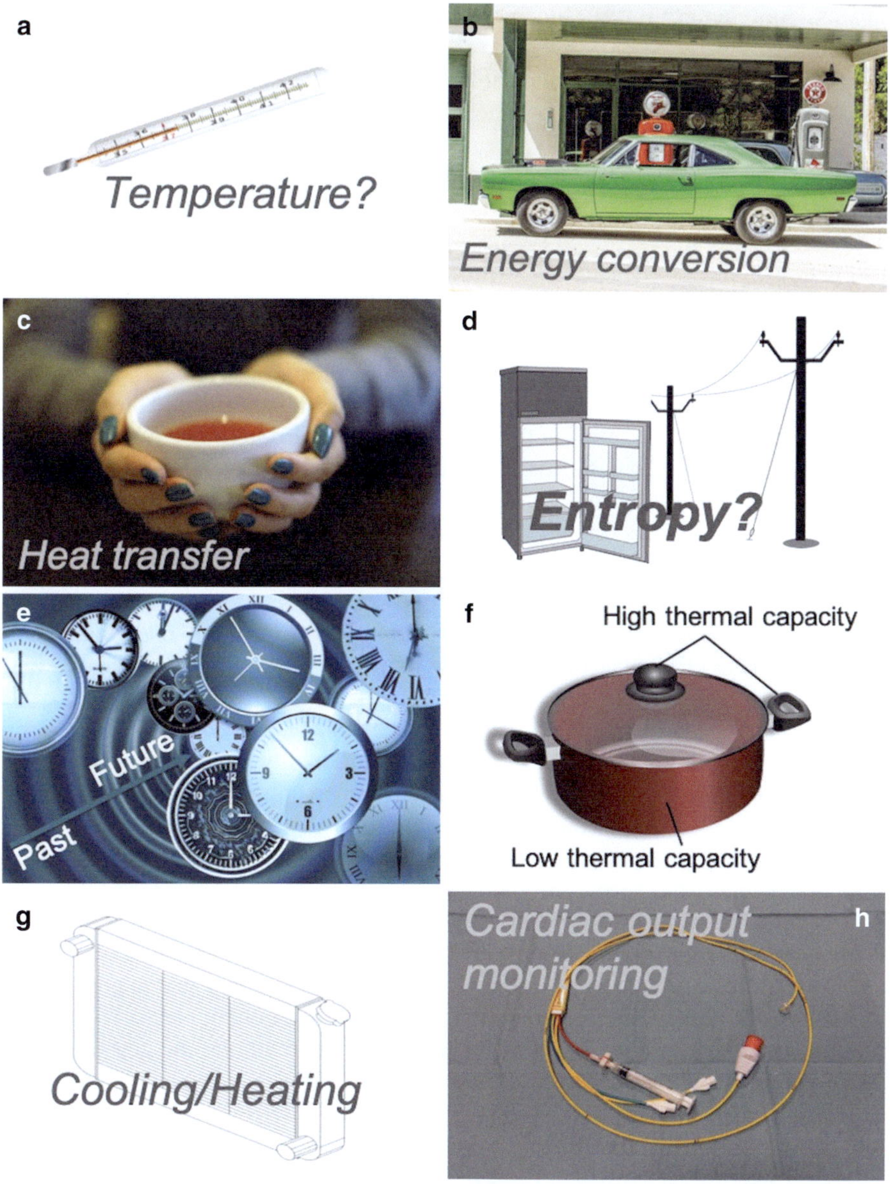

Fig. 10.1 Thermodynamics and its laws in everyday life, including that of the operating room/intensive care unit (**a**–**h**: see text)

Thanks to the zeroth law of thermodynamics, we can say that the objects X and Z have an equal temperature without putting them into contact, using the object Y as a *thermoscope* (i.e., something with a measurable property which changes with temperature). A thermoscope can be turned into a *thermometer* after calibration, i.e., assigning an arbitrary "value" of temperature to certain reproducible thermal phenomena, e.g., 0 °C (32 °F) to the freezing point of water and (about) 100 °C (212 °F) to its boiling point at standard atmospheric pressure.

10.1.2 The First Law of Thermodynamics: It All Adds Up!

The *first law of thermodynamics* expresses the principle of energy conservation for a system which is not isolated (i.e., can exchange heat and *work* with the surrounding environment). It is usually formulated as follows:

$$\Delta U = Q - W \tag{10.1}$$

that is, the change in the *internal energy* ΔU of a system is equal to the difference between the net heat Q supplied to the system by its surrounding environment and the net work W (see Chap. 6, Box 6.1) done by the system itself. Alternatively, the work done *on* the system can be considered. In this case, the first law of thermodynamics can be written as follows:

$$\Delta U = Q + W \tag{10.2}$$

In other words, energy (e.g., heat) can be transferred from one body to another, or from a system to its surroundings and vice versa. It can also be converted into other forms of energy, such as mechanical work (which, as mentioned in Box 6.1, is "energy transfer" by means of a force). Indeed, mechanical work and heat are fully equivalent to each other (with an adequate proportionality coefficient); that is, the same change in the internal energy of a system can be obtained by either providing a certain quantity of heat to the system or doing some external work on it.

The only thing that energy cannot do is to disappear into nothing (or be created from nowhere).

Thus, if a car engine has a *thermal efficiency* of, say, 30% (i.e., only the 30% of the *potential energy* which is stored in the chemical bonds of the fuel is converted into useful mechanical work), the remaining 70% of the energy that is provided to it does not disappear, but is converted into other forms of energy (mostly heat, but also the work which is needed to overcome the friction among the gears, and the *sound energy* which produces the engine noise).

10.1.3 In which Direction, Please? The Second Law of Thermodynamics

The second law of thermodynamics can be formulated in many ways. In its earliest form, which is known as *Kelvin statement*, it states that it is not possible to realize a process whose only result is to completely convert the heat absorbed from a thermal reservoir into work. In other words, a perfect engine (with a 100% thermal efficiency) cannot exist. Another way of expressing the second law of thermodynamics is the so-called *Clausius statement*: no process is possible whose only result is the transfer of heat from a colder body to a hotter one.

In order to describe mathematically the second law of thermodynamics (and to show that the two above-mentioned statements say exactly the same thing), we need to introduce a new thermodynamic quantity: *entropy*.

Entropy is a measure of the "disorder" of a system and, at the same time, of the ability of that system to do work. In other words, entropy is the thermal energy per unit temperature which is *not* available for doing work; accordingly, it represents a measure of the "quality" of energy that is stored in a system. Entropy is, as physicists say, a *state function*, i.e., its value for a given state of the system is independent of how that state has been reached.

A just bought pack of playing cards (in which the cards are sorted in ascending order and separated by suit) has low entropy. When the cards are shuffled, entropy increases. The entropy of a charged battery is low, but it is much higher when the battery is exhausted (and can no longer do work, e.g., move the hands of a clock). These and other revealing (and funny) examples can be found in two beautiful books by the physicist Jim Al-Khalili [2, 3].

When some heat is absorbed from the surrounding environment, or released to it, by a system, the *change in entropy* ΔS of such a system can be calculated, roughly (i.e., without resorting to the integral calculus), as follows:

$$\Delta S = \frac{\Delta Q}{T} \tag{10.3}$$

where ΔQ is the energy transferred (as heat) between the system and the environment and T is the average temperature (in kelvins) at which the transfer occurred.

Hence, the same quantity of heat provided to a system will produce a greater increase in its entropy if temperature is lower, and a smaller increase in its entropy if temperature is higher (i.e., the system is already rather "chaotic"). Using a really enlightening analogy, Peter Atkins says that a sneeze causes a much higher increase of "entropy" in a silent library than in a busy road [1].

Now we are ready to formulate the second law of thermodynamics in equation form:

$$\Delta S \geq 0 \tag{10.4}$$

that is, entropy of a closed system can only increase (as a result of an *irreversible* process) or remain constant (following a *reversible* process), but *it can never decrease*.

If we consider the Universe as a closed system, we could say that the entropy of the Universe always increases during a spontaneous transformation. However, the entropy of any system which is located within the Universe (and is not isolated from it) can certainly decrease, provided that entropy increases somewhere else. If you hold a cup of hot tea between hands (as in Fig. 10.1c), heat will be transferred from the tea cup to your hands because this is the "direction" that the increase in entropy imposes (i.e., from the hotter body to the colder one), and *nothing else happens* (as the Clausius statement specifies). The fridge in your kitchen (in which total entropy decreases since heat is continuously removed from warm food and beverages and transferred to the warmer environment) does not violate the second law of thermodynamics; in fact, entropy is increasing, at the same time, in a power plant, so that the entropy of the system "fridge + power plant" increases (Fig. 10.1d).

Ultimately, the second law of thermodynamics sets the direction in which time flows, namely that of the increase in the entropy of Universe, i.e., of the reduction in the "quality" of its total energy. Accordingly, it is also known as the "arrow of time." At the moment, this is the best way to distinguish past from future!

10.1.3.1 Second Law of Thermodynamics, Metabolic Heat Dissipation, and the Zero-Heat-Flux Thermometer

Moving from the immensity of the Universe to our small lives (both as human beings in general and as anesthesiologists or intensivists), it is precisely the second law of thermodynamics which imposes that the body *core temperature* (i.e., roughly, the temperature of trunk and head) must be higher (usually by about 2–4 °C at hospital room temperature) than peripheral temperature (limbs), and that, in turn, the latter must be higher than ambient temperature (see Fig. 10.2); in fact, according to the second law of thermodynamics, a temperature gradient is needed for metabolic heat to be transferred from the core to peripheral tissues and then to room air [4]. Hence, as common sense teaches us, a blanket (which provides thermal insulation, preventing or slowing down the heat flow along the temperature gradient) can be used to keep a human body warm. Someone who has never studied physics might think (instinctively) to warming up a blood bag, or anything else that is colder than ambient air and which does not produce heat by itself, by covering it with a blanket (you won't believe it, but once, a long time ago, I happened to see even this); of course, the only result may be (perhaps) to keep it cold for longer!

The second law of thermodynamics also allows to monitor *core* temperature in the operating room and in the intensive care unit (ICU) using a particular skin thermometer, called *zero-heat-flux (ZHF) thermometer* [5–7]. Thanks to a system consisting of two thermometers separated by an insulator and covered by a servo-controlled heater which maintains the two thermometers at the same temperature (thus "interrupting" the core-to-ambient temperature gradient), the skin area under the ZHF probe is perfectly insulated from the outside and, at equilibrium, a column of tissues is created all having (approximately) the same temperature, which approaches core temperature. In fact, according to the second law of thermodynamics, heat will "flow" from the core to the peripheral compartment (as long as there is a temperature gradient), but not from the latter to the ambient (since there is no temperature gradient across the insulator) [5, 6]. If the skin is close enough to the core thermal compartment (as at the forehead level), the temperature measured by the ZHF thermometer may represent an acceptable estimation of the core

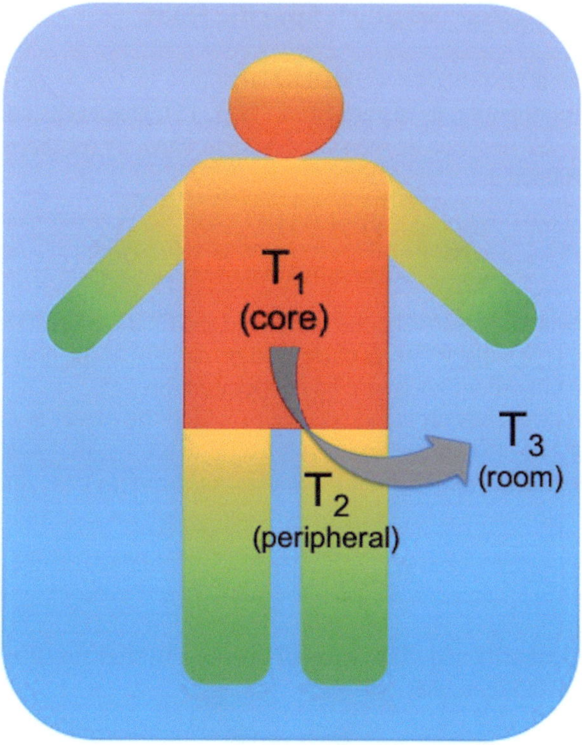

Fig. 10.2 According to the second law of thermodynamics, core body temperature (T_1) must be higher than peripheral body temperature (T_2) which, in turn, must be higher than ambient temperature (T_3) for the metabolic heat to be dissipated (arrow). In fact, heat can be transferred from the core thermal compartment (head and trunk) to the peripheral thermal compartment of the body (limbs), and from the latter to the environment, only in the presence of a temperature gradient along this "path"

temperature monitored with more invasive devices such as esophageal or nasopharyngeal thermistors/thermocouples or the pulmonary artery catheter thermistor (see below and Chap. 11) [5, 7].

10.1.4 There Is Also a Third Law of Thermodynamics (Just to Know)

The third law of thermodynamics (also referred to as *Nernst's theorem*) states something like this: it is not possible to cool a body up to the *absolute zero* in a finite number of operations. Most physicists say it is not even a "law" of thermodynamics, but simply something which follows from the second law. Furthermore, it deals with extremely low temperatures (absolute zero is the zero on the kelvin scale, i.e., −273.16 °C!); hence, we will not care about it here.

10.2 More or Less "Greedy": Specific Heat

When some heat is absorbed or released by an object, its temperature varies (unless a state change is ongoing, as discussed in Chap. 15). However, the extent of temperature change ΔT depends on the so-called *heat capacity* (or *thermal capacity*) C of the object, defined as:

$$C = \frac{Q}{T_1 - T_0} = \frac{Q}{\Delta T} \text{ or } Q = C\,\Delta T \tag{10.5}$$

where T_0 and T_1 are, respectively, the initial and the final temperature of the object. In other words, heat capacity of an object can be defined as the heat which needs to be provided to it in order to increase its temperature by 1 °C.

Heat capacity depends on the substance of which the object is made and on its mass (see also Chap. 15). Accordingly, it is useful to refer to the *specific heat* c (also known as *specific heat capacity* or *specific thermal capacity*), i.e., the heat capacity per unit mass:

$$c = \frac{C}{m} \tag{10.6}$$

where C is heat capacity and m the mass of the object. After substituting for C from Eq. (10.6), Eq. (10.5) becomes:

$$Q = c\,m\,\Delta T \tag{10.7}$$

that is, specific heat c is the heat which is needed in order to raise 1 unit mass of a substance by 1 unit temperature.

For example, pans and pots are usually made of materials with low specific heat, so that a small amount of heat is needed to increase their temperature enough to cook your meals. Conversely, their handles should be preferably made of materials with high specific heat, as to not get burned when you touch them (Fig. 10.1f).

Water has a very high specific heat (e.g., about 5 times that of aluminum and 30 times that of mercury), i.e., it absorbs a great amount of heat with little increases in its temperature; thanks to this property, it is used in car radiators as a cooling fluid. In other words, water can "carry" a lot of heat, and it releases great amounts of heat with small decreases in its temperature. This is one of the reasons why, in the absence of the sea or large lakes (as in the desert), there is a wide thermal excursion between night and day; in fact, large masses of water can store a lot of heat during the sunny hours, which they "return" overnight (so that air temperature does not drop too much). For the same reasons, water is used in home central heating systems as well as in the heat exchanger of the heart-lung machine used for cardiopulmonary bypass (CPB) [8].

We will encounter the concept of specific heat again when addressing vaporizers for volatile anesthetic agents (Chap. 15), but it was essential to discuss it here before moving to the next section.

10.3 Measuring Cardiac Output by Thermodilution

We all know that we can measure cardiac output (CO) with a pulmonary artery catheter (PAC) by means of the so-called *thermodilution* method [9–11]. A given volume of iced or room-temperature fluid (e.g., saline) is injected into the right atrium through the proximal lumen of the PAC, and the resulting change in blood temperature is measured in the pulmonary artery by a thermistor (see Chap. 11) which is located at the tip of the PAC. Cardiac output is inversely proportional to the area under the curve of pulmonary artery blood temperature change over time (the so-called *thermodilution curve*), according to the modified *Stewart-Hamilton equation*:

$$\mathrm{CO} = V_\mathrm{i}\, K_1 \frac{\left(T_\mathrm{b} - T_\mathrm{i}\right)}{\int_0^{\infty} \Delta T_\mathrm{b}\left(t\right)\mathrm{d}t} \frac{d_\mathrm{i}\, c_\mathrm{i}}{d_\mathrm{b}\, c_\mathrm{b}} \tag{10.8}$$

where V_i is the volume of injectate; T_b and T_i are baseline blood temperature and injectate temperature, respectively; d_i and c_i are, respectively, the density (or specific gravity, see Chap. 9) and the specific heat of the injectate; d_b and c_b are, respectively, the density and the specific heat of blood; K_1 is a computation constant which accounts for the heat exchange in transit, the catheter dead space, the injection time, and the unit conversion; and $\int_0^{\infty} \Delta T_\mathrm{b}(t)\, \mathrm{d}t$ is the integral (see Chap. 1) of temperature change over time, i.e., the area under the thermodilution curve.

To make it less "frightening" (anyhow, it is the monitor software that does the calculation!), Eq. (10.8) is usually reported in the following short form:

$$\mathrm{CO} = \frac{\left(T_\mathrm{b} - T_\mathrm{i}\right)K}{\int_0^{\infty} \Delta T_\mathrm{b}\left(t\right)\mathrm{d}t} \tag{10.9}$$

where K is a "comprehensive" constant defined as:

$$K = V_\mathrm{i}\ K_1 \frac{d_\mathrm{i}\, c_\mathrm{i}}{d_\mathrm{b}\, c_\mathrm{b}} = V_\mathrm{i}\ K_1\, K_2 \tag{10.10}$$

with $K_2 = (d_\mathrm{i}\, c_\mathrm{i})/(d_\mathrm{b}\, c_\mathrm{b})$.

While the Stewart-Hamilton equation, in any of its forms, can be found in every anesthesia or critical care textbook, its "origin" is rather shrouded in mystery (except for a few vague references to the fact that it follows from the laws of thermodynamics or from a "thermal energy balance" [9, 11]). Let's try, then, to give a sense (for anesthesiologists and intensivists) to what we discussed so far.

10.3.1 It's Just Thermodynamics, Beauty!

When a fluid bolus is injected into the right atrium, some heat is transferred from blood to the injectate (according to the second law of thermodynamics), leading to a reduction in temperature. According to Eq. (10.7), the amount of heat transferred can be calculated as follows:

$$Q_1 = c_i\, m_i\, \Delta T = c_i\, m_i \left(T_b - T_i\right) \tag{10.11}$$

where c_i, m_i, and T_i are, respectively, the specific heat, mass, and temperature of the injectate, and T_b is the baseline blood temperature.

Remembering the definition of density (see Chap. 9) and substituting by Eq. (9.1), Eq. (10.11) becomes:

$$Q_1 = c_i\, d_i\, V_i \left(T_b - T_i\right) \tag{10.12}$$

where d_i and V_i are the density and the volume of injectate, respectively.

Afterwards, *due to blood flow*, further heat exchange occurs, leading to the observed change in the pulmonary artery blood temperature over time. As above, we can apply Eq. (10.7) to calculate such heat transfer (whose "rate" evidently depends on cardiac output):

$$Q_2 = \Sigma c_b\, m_b \left(T_b - T_{(t)}\right) = \Sigma c_b d_b\, V_b \left(T_b - T_{(t)}\right) = \Sigma c_b d_b\, V_b\, \Delta T_{(t)} \tag{10.13}$$

where c_b, d_b, V_b, and T_b are, respectively, specific heat, density, volume, and (baseline) temperature of blood, $T_{(t)}$ is pulmonary artery blood temperature "as it changes over time," $\Delta T_{(t)}$ is the change in pulmonary artery blood temperature over time (t), and the symbol Σ means that the total heat transfer is the sum of heats transferred for any value of pulmonary artery blood temperature over time (this is a not too rigorous notation from a mathematical standpoint, but I hope the concept is clear).

As mentioned in Chap. 7, cardiac output (CO) is the *volume flow rate* V_b/t (volume of blood per unit time t). Hence, after substituting by $V_b = \text{CO}\ t$, Eq. (10.13) becomes:

$$Q_2 = \Sigma c_b\, d_b\, \text{CO}\, t\, \Delta T_{(t)} \tag{10.14}$$

According to the first law of thermodynamics (Eq. 10.2), the change in the internal energy of the system (in this case, the right heart and the pulmonary artery or, more generally, the human body) is equal to the heat transferred (considering the work done on the system as negligible). Hence, remembering that "energy is conserved" (see Chap. 7, Box 7.1) and assuming that blood and the injectate are mixed completely and instantaneously and that no other heat exchanges occur (and, accordingly, that pulmonary artery blood temperature remains constant except for the changes induced by the fluid bolus), it must be $Q_1 = Q_2$. From Eqs. (10.12) and (10.14), it is therefore:

$$c_i d_i V_i \left(T_b - T_i\right) = \Sigma c_b\, d_b\, \text{CO}\, t\, \Delta T_{(t)} \tag{10.15}$$

or, rearranging a bit:

$$\text{CO} = V_i \frac{\left(T_b - T_i\right)}{\Sigma \Delta T_{(t)} t} \frac{c_i\, d_i}{c_b\, d_b} \tag{10.16}$$

As you can see, Eq. (10.16) closely resembles Eq. (10.8) (modified Stewart-Hamilton equation) and becomes identical to it remembering that the quantity roughly indicated as $\sum \Delta T_{(t)}\, t$ is more correctly described by the integral of the thermodilution curve $\int_0^{\infty} \Delta\, T_b(t)\, dt$ (see Chap. 1, Sect. 1.4.2), and adding an adequate computational constant K_1.

A thorough knowledge of the modified Stewart-Hamilton equation and of its "thermodynamic background" may help us to not blindly trust our thermodilution CO measurements: the number displayed on the monitor, although clinically useful, is not necessarily the "true" cardiac output. In fact, given its large number of assumptions, the thermodilution method has several sources of error [9–12]. For example, we assumed that there are no other heat exchanges apart from the injectate bolus. However, pulmonary artery blood temperature may vary due to physiological fluctuations with respiration, rapid intravenous fluid administration, and the so-called baseline drift soon after hypothermic CPB [9–11]. Moreover, the accuracy of the volume of injectate and, most importantly, of the measurement of its temperature, are pivotal. Finally, as shown, the thermodilution CO equation includes the specific heat of blood (c_b) as a constant: however, this blood property is not indeed constant, but it varies slightly with the hematocrit [11, 13].

References

1. Atkins P. Four laws that drive the universe. New York: Oxford University Press; 2007.
2. Al-Khalili J. Black holes, wormholes and time machines. 2nd ed. Boca Raton: CRC Press; 2011.
3. Al-Khalili J. Paradox. The nine greatest enigmas in science. Broadway Paperbacks; 2012.
4. Sessler DI. Perioperative thermoregulation and heat balance. Lancet. 2016;387(10038):2655–64.
5. Sessler DI. Perioperative temperature monitoring. Anesthesiology. 2021;134(1):111–8.
6. Eshraghi Y, Nasr V, Parra-Sanchez I, et al. An evaluation of a zero-heat-flux cutaneous thermometer in cardiac surgical patients. Anesth Analg. 2014;119(3):543–9.
7. Sastre JA, Pascual MJ, López T. Evaluation of the novel non-invasive zero-heat-flux Tcore™ thermometer in cardiac surgical patients. J Clin Monit Comput. 2019;33(1):165–72.
8. Hines MH. Principles of oxygenator function: gas exchange, heat transfer, and operation. In: Gravlee GP, Davis RF, Hammon JW, Kussman BD, editors. Cardiopulmonary bypass and mechanical support. Principles and practice. 4th ed. Philadelphia: Wolters Kluwer; 2016. p. 95–108.
9. Moise SF, Sinclair CJ, Scott DH. Pulmonary artery blood temperature and the measurement of cardiac output by thermodilution. Anaesthesia. 2002;57(6):562–6.
10. Mark JB, Slaughter TF. Cardiovascular monitoring. In: Miller RD, editor. Miller's anesthesia. 6th ed. Philadelphia, PA: Elsevier Churchill Livingstone; 2005. p. 1265–362.
11. Szocik JF, Barker SJ, Tremper KK. Fundamental principles of monitoring instrumentation. In: Miller RD, editor. Miller's anesthesia. 6th ed. Philadelphia, PA: Elsevier Churchill Livingstone; 2005. p. 1191–225.
12. Nishikawa T, Dohi S. Errors in the measurement of cardiac output by thermodilution. Can J Anaesth. 1993;40(2):142–53.
13. Mendlowitz M. The specific heat of human blood. Science. 1948;107(2769):97–8.

Electric Current, Resistance, Circuits, Thermoelectric Effect: Platelet Aggregometry, Pressure Transducers, and Temperature Monitoring

11

Contents

Where's Physics

Daily life	Light bulbs, home appliances, and everything that runs on electricity Temperature setting at home
Physics involved	Electricity (charge, electric field, electric potential, electric current) Electrical resistance (Ohm's law) Thermoelectric effect (Seebeck effect) Circuit elements (resistors, thermistors, thermocouples) Kirchhoff's laws
Clinical practice	Temperature monitoring (thermistors and thermocouples) Platelet function testing (multiple electrode aggregometry) Invasive pressure transducers (Wheatstone bridge)

A. Pisano, *Physics for Anesthesiologists and Intensivists*,
https://doi.org/10.1007/978-3-030-72047-6_11

11.1 Electricity: A (Very) Concise Introduction

In this and the next chapter, we need to talk about *electricity*. This part of physics, in fact, includes the description of several quantities, phenomena, and devices which are closely related to what we deal with every day as anesthesiologists or intensivists. For example, the measurement of a change in the *electrical resistance* of "something" is involved in invasive pressure monitoring, temperature monitoring, cardiac output estimation, and platelet function testing. Furthermore, body temperature can be measured in clinical practice also using a circuit element, called thermocouple, which relies on a particular *thermoelectric effect*. Finally, in Chap. 12 we will talk about *capacitors*: as we'll see, a capacitor is a component designed to exploit its *capacitance* in electrical or electronic devices and is the essential constituent of a defibrillator.

In physics textbooks, these notions are addressed in hundreds of pages together with many others which are often "preparatory" and cannot be ignored to understand the former. Of course, a systematic and thorough discussion of this matter is beyond our scope here. However, in the full spirit of this book, which is certainly to cover only physics topics with strict implications for the practice of anesthesia and critical care, but in the most "immersive" way possible, it is necessary to say at least something about charge, electric field, and electric potential before discussing about current, resistance, circuits, thermoelectric effect, capacitance (the latter in Chap. 12), and their clinical applications.

11.1.1 Charge

We all know that there are two types of electric charge (usually symbolized q): *positive* ($+q$) and *negative* ($-q$). These two names were coined by Benjamin Franklin (one of the Founding Fathers of the United States), and are completely arbitrary. That is, the only thing that distinguishes a positive charge from a negative charge is that two objects which have opposite charges attract each other, while two objects with a charge of the same sign repel each other (we all know this too).

In particular, according to *Coulomb's law*, two charged particles with charges q_1 and q_2, respectively, exert on each other an *electrostatic* (or *electric*) *force* of magnitude:

$$F = \frac{1}{4\pi\varepsilon_0}\frac{|q_1||q_2|}{r^2} \tag{11.1}$$

or, in a simpler form,

$$F = k\frac{|q_1||q_2|}{r^2} \tag{11.2}$$

where r is the distance between the two particles, ε_0 is a quantity called *permittivity constant*, $k = 1/4\pi\varepsilon_0$ is the so-called *electrostatic constant* (or *Coulomb constant*), and the symbol | | indicates that the charges are expresses as absolute values (i.e., without the plus or minus sign). Of course, the force is attractive for opposite charges and repulsive for charges of the same sign; moreover, as evident from Eqs. (11.1) or (11.2), it becomes weaker as the distance between the two particles increases.

But, what electric charge exactly *is*? It can be said that charge is the physical property of a particle which causes it to be subject to a *force* when placed in an *electric field* (or, more generally, in an electromagnetic field). This is absolutely correct, but does not help much. In fact, as discussed below, an electric field is the force field which develops in the space surrounding any electric charge! Doesn't that remind you of the chicken and egg affair? We must be content with considering electric charge as an intrinsic property of the particles of which we and everything around us are made.

Charge is a *quantized* property: it can only assume values which are integer multiples of an *elementary charge e*, namely that of the electron ($-e$) or of the proton ($+e$) (see Box 11.1). That is, any charge which can be found in nature has a value:

$$q = \pm ne \tag{11.3}$$

with $n = 1, 2, 3, \ldots$ Ultimately, a charged object simply contains more electrons than protons (negative "excess charge"), or vice versa (positive "excess charge"), while an object which contains the same number of electrons and protons is said to be *electrically neutral.*

As mentioned in Chap. 7 (Box 7.1), charge is one of the physical quantities which, in an isolated system, are always conserved; that is, to use the simple words of Richard Feynman (one of the best physicists ever), the sum of all positive charges minus that of all negative charges contained in an isolated system, always gives the same result (*law of conservation of electric charge*) [1].

Box 11.1: Just out of Curiosity: Elementary Charge and Elementary Particles

The *elementary charge* has the following value:

$$e \cong 1.6 \times 10^{-19}\ \mathrm{C}$$

where C stands for *Coulomb*, the unit of charge in the International System of Units (we have to talk about electric current before we can define it).

As mentioned, *e* is the magnitude of the charge of the electron ($-e$) and of the proton ($+e$).

One might think that the elementary charge is, simply, the charge of elementary particles; this is only partly true. In fact, according to the Standard Model for particle physics, while the electron is an elementary particle, the proton *is not*. Indeed, a proton is made up of two *up* quarks (which have a charge of $+\frac{2}{3}e$) and one *down* quark (which have a charge of $-\frac{1}{3}e$);

accordingly, its charge is $+\frac{2}{3}e+\frac{2}{3}e-\frac{1}{3}e=+e$ (for your knowledge, a neutron is instead made up of one *up* quark and two *down* quarks, so its charge is $+\frac{2}{3}e-\frac{1}{3}e-\frac{1}{3}e=0$). As you can see, there are elementary particles with a smaller charge than *e*. However, no one has ever seen a quark alone; hence, our "elementary charge" (i.e., the smallest charge available in nature) remains $\pm e$.

11.1.2 Electric Field

Any electric charge (or any set of electric charges) creates an alteration of the space all around so that any other charged particle which is located in that space is subject to an electrostatic (or electric) force, whose magnitude and direction depend on the position of the particle with respect to the "source" charge(s). This "alteration of the space" is called *electric field*. Let's try to define it better.

In general, the concept of "field" in physics refers to a function which assigns a certain value of "something" at every point in space. In particular, electric field is a *vector field*. In fact, as mentioned, it describes the *force* which acts on a given charged particle at every point in space and, as discussed in Chap. 13, force is a vector quantity (i.e., you need to specify its *direction* in addition to its "magnitude" to define it).

Let's consider, for example, the electric field created by a point (i.e., dimensionless) charge (Fig. 11.1). At any point around this source charge, a positive test charge q_0 will be subject to a certain electrostatic force $\vec{F}$ (the arrow indicates that it is a vector). If we measure this force, we can define the *electric field vector* at that point as:

$$\vec{E}=\frac{\vec{F}}{q_0} \tag{11.4}$$

In practice, the electric field represents the distribution of the electric field vectors in the space around the source charge. As shown in Fig. 11.1, the electric field vectors are directed toward the source charge if it is negative, and away from the source charge if it is positive; moreover, the magnitude of $\vec{E}$ decreases with the distance from the source charge (just remember Coulomb's law). According to Eq. (11.4), any particle with charge q which is located at a point P where the electric field is $\vec{E}$ experiences a force $\vec{F}=q\vec{E}$ exerted by the field. This force has the same direction of $\vec{E}$ if q is positive, while it is directed opposite to $\vec{E}$ if q is negative (see Fig. 11.1).

If you remember what we said in Chapters 6 and 7 about work and energy (see, in particular, Box 6.1), namely that any force which makes an object move is doing a work, and that work and energy are practically the same thing, it is evident that to

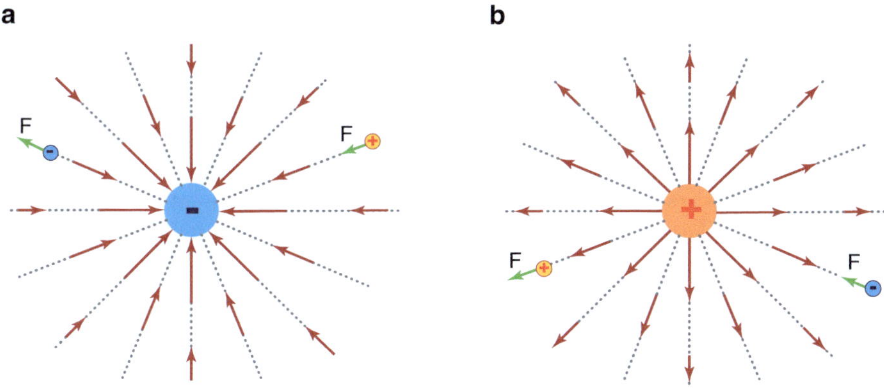

Fig. 11.1 The electric field generated by a negative (**a**) or positive (**b**) point charge (source charge). The magnitude of the electric field vectors (red arrows) decreases as the distance from the source charge increases. Of course, there are infinite "field lines" (dashed lines) between those drawn, and electric field vectors of different magnitude at any point along these lines. Regardless of the source charge sign, the electrostatic force *F* (green arrows) acting on a charge which is located at any point in the electric field has the same direction of the electric field vector in that point for a positive charge, while it is directed opposite to the electric field vector for a negative charge

any electric field must be associated some energy, particularly a *potential energy*; in fact, the electric field has the "potential" to do some work on a charged particle positioned inside it.

We can define the *energy density* u_E, i.e., the potential energy per unit volume, of an electric field of magnitude *E* (in vacuum) as:

$$u_E = \frac{1}{2}\varepsilon_0 E^2 \tag{11.5}$$

where ε_0 is the vacuum permittivity constant. As it can be expected, the higher the magnitude of an electric field, the greater the energy stored in it.

This definition will come in handy when discussing about capacitors (and defibrillators) in Chap. 12.

11.1.3 Electric Potential

Each of us has sometimes gone out to buy a 1.5-volt battery; all people know that the electric mains (or power supply) in his/her home is 220 volts (or 230, 120, etc., depending on the Country); finally, everyone knows it's best to stay away from a sign that reads: "warning, 12000 volts"! This is a case in which the unit is much more known than the quantity that it measures. Maybe, someone even remembers that *volt* (V) is the unit of *electric potential*, most often expressed as a *difference* in the electric potential between two points, i.e., "potential difference" (or *voltage*), rather than as an "absolute" value. However, if you ask anyone who hasn't studied physics thoroughly what electric potential is, the matter becomes somewhat complicated. In fact, there are no other words than the (strictly "physical") following to define it: the electric potential V at any point P in an electric field is the work W done by the electric field forces to move a charged particle with (positive) unit charge from the point P to an infinite distance or, which is the same thing, the *electric potential energy U* which is stored in the electric field + charged particle system when a charged particle, with any charge q, is moved from an infinite distance to P, divided by q. In equation form:

$$V = \frac{-W}{q} = \frac{U}{q} \tag{11.6}$$

where the minus sign indicates that the electric potential energy which is stored in the system when a charge q is moved from an infinite distance to the point P is equal to the work that the electric field forces must do to make the charge perform the inverse path (i.e., from P to an infinite distance).

In practice, as mentioned, we usually refer to the *potential difference* between two points A and B, which represents the work done by the electric field forces to move the (positive) unit charge from A to B. If the charge displaced from A to B has any value q, then this work is evidently:

$$W_{\mathrm{AB}} = q\left(V_{\mathrm{A}} - V_{\mathrm{B}}\right) = q\Delta V \tag{11.7}$$

where $\Delta V = V_{\mathrm{A}}$-$V_{\mathrm{B}}$ is the potential difference between the points A and B (*voltage*). It comes from Eq. (11.6) that the electric potential unit is the joule (i.e., the unit of work/energy) per coulomb (i.e., the unit of charge, see below), which is called *volt* (V), after Alessandro Volta, the Italian scientist who invented the electric battery between the end of the eighteenth and the beginning of the nineteenth century. According to Eq. (11.7), there is a potential difference of 1 V between two points A and B in an electric field when the electric field forces do a work of 1 joule (J) to move a positive charge of 1 coulomb (C) from A to B.

Unlike the electric field, which, as mentioned, is a vector distribution, the electric potential is a *scalar* quantity, i.e., a number is enough to define it (see Chap. 13). Hence, in simple terms, electric potential provides us with a convenient way to "understand" how electric charges move within an electric field. In fact, since the electric field is a *conservative field* (i.e., the electrostatic forces "generated" by the electric field are conservative forces), which means that it must obey the *law of conservation of mechanical energy* (see Chap. 7, Box 7.1), the kinetic energy of a positive charge must decrease as it moves toward higher electric potential. In fact, according to Eq. (11.6), a charged particle with charge q which is located at a point where the electric potential V has an electric potential energy $U = qV$. Hence, when the particle moves to a point with higher electric potential, its potential energy increases and, since mechanical energy is the sum of kinetic energy and potential energy, its kinetic energy must decrease in order to keep mechanical energy unchanged. In practice, a positive charge slows down as it moves toward higher electric potential (or through a *positive* potential difference) and accelerates as it moves toward lower electric potential (or through a *negative* potential difference). The opposite occurs to a negative charge. And this brings us straight to the next section.

11.2 Electric Current, Resistance, and Circuits

If a potential difference is applied to the ends of a conductor, i.e., any material in which charges are free to move, a "flow of charges" will develop through it (as you can easily realize based on what discussed in Sect. 11.1.3). We call this flow of charges an *electric current*. In particular, we can define the current (or current *intensity*) i through a conductor as the charge q which passes any cross section of the conductor per unit time:

$$i = \frac{q}{t} \tag{11.8}$$

or, for the lovers of mathematical rigor,

$$i = \frac{\mathrm{d}q}{\mathrm{d}t} \tag{11.9}$$

i.e., the *derivative* of electric charge by time (we have already seen in Chap. 1 the reason why a derivative provides a more general and rigorous definition, as compared with a ratio, of quantities such as velocity, gradient, and electric current; however, a ratio is equally correct if these quantities are constant over time, or uniform in space in the case of gradients).

According to Eqs. (11.8) and (11.9), the unit of current in the International System of Units (SI) is the coulomb (i.e., the SI unit of charge) per second (C/s), which is called ampere (A). However, for a matter of practicality and precision, the coulomb is derived, in turn, from the definition of current, and this is why we had to wait until now before defining it (as mentioned in Box 11.1): a charge of 1 C is the charge carried in 1 second by a current of 1 A.

As stated above, for there to be a current through a conductor there must be a potential difference between its ends. In particular, according to the so-called *Ohm's law*, the current through a circuit element is *directly proportional* to the potential difference across it:

$$i = \frac{\Delta V}{R} \tag{11.10}$$

where R is the *electrical resistance* of that circuit element, which is measured in volt per ampere, or Ohm (Ω). In practice, the higher the potential difference across a circuit element, the more intense is the current through it, while applying the same potential difference across two circuit elements with different resistance will give rise to a weaker current through the element with higher resistance and to a more intense current through that with lower resistance.

Electrical resistance depends on the *resistivity* of the material of which the conductor is made (which is a property of any material which expresses, roughly, the "reluctance" of its electrons to move when exposed to an electric field) as well as on its shape and dimensions. For example, for a homogeneous conductor with uniform cross section such as a copper wire, resistance can be calculated as follows:

$$R = \rho \frac{L}{A} \tag{11.11}$$

where L and A are the length and the cross-sectional area of the conductor, respectively, and ρ is the resistivity (in this case, of copper).

Note that Ohm's law is only valid within certain values of the electric field and for homogeneous materials whose resistivity does not change with the potential difference applied. These materials are called "ohmic" (since they obey Ohm's law). Ohmic materials used in common *electric circuits* can be classified as *wires* (with very low resistance), *resistors* (i.e., elements designed to provide a specific resistance in a circuit), and *insulators*, which have a very high resistance (tending to infinity in an "ideal" insulator).

Figure 11.2 shows the *circuit diagram* of the simplest electric circuit, which consists of a resistor connected by wires to a battery (namely the source of the potential difference). Although in metals (e.g., those of which electric wires and other circuit elements are made) the charges which are "free to flow" and that, in practice, form the electric current, are represented by electrons, the direction of current is, by convention, that of "virtual" positive charges which would move in the opposite direction.

Current in an electric circuit is regulated by the two *Kirchhoff's laws* (Fig. 11.3). Due to conservation of charge (see Sect. 11.1.1), the total current entering a circuit junction must be equal to that leaving it (*Kirchhoff's current law*, also referred to as

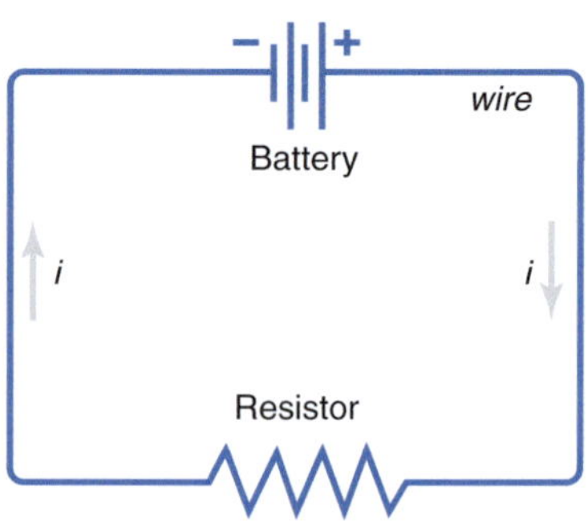

Fig. 11.2 The diagram of the simplest electric circuit consisting of a resistor connected by wires to a battery. As shown, the current direction is, by convention, that in which positive charges would move, although the charges which indeed move (in the opposite direction), i.e., electrons, are negative

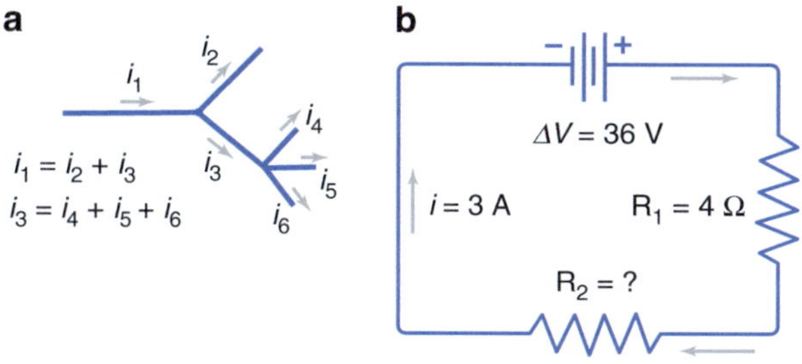

Fig. 11.3 Kirchhoff's laws. (**a**) According to the Kirchhoff's *current* (or junction) law, the total current (i) entering a junction in an electric circuit must be equal to the total current leaving that junction. (**b**) The Kirchhoff's *voltage* (or loop) law states that the algebraic sum of the potential differences across all the circuit elements in any loop of a circuit must be zero. This allows to calculate the potential difference across the resistor with unknown resistance R_2 (and the value of R_2 itself by knowing the current intensity) in the depicted circuit according to the potential differences across the other two elements, i.e., the battery and the other resistor (see text). ΔV Potential difference; R_1 and R_2 Resistances of the two resistors

Kirchhoff's junction law or *Kirchhoff's junction rule*). Moreover, due to conservation of energy (see Chap. 7), when a charge moves along a closed path (i.e., a loop), its potential energy does not change; accordingly, the algebraic sum (i.e., a sum which takes into account the sign of the operands) of the potential differences across all the circuit elements in any loop of a circuit must be zero (*Kirchhoff's voltage law* or *Kirchhoff's loop rule*). For example, in a circuit loop such as that shown in Fig. 11.3b, the unknown resistance R_2 can be calculated according to this law. In this case, the battery has a positive potential difference (voltage) $\Delta V = 36$ V, which generates a current $i = 3$ A in the circuit, while the two resistors (with resistance R_1 and R_2, respectively) cause a potential difference drop (so we have to put the minus sign) of magnitude $i\ R_1$ and $i\ R_2$ (just take a look to Eq. (11.10)). Hence, according to Kirchhoff's voltage law, it must be:

$$\Delta V - i R_1 - i R_2 = 0 \tag{11.12}$$

or

$$\Delta V = i R_1 + i R_2 \tag{11.13}$$

Doing the math: 36 V = (3A x 4Ω) + (3A x R_2) and, resolving, $R_2 = 8\ \Omega$. If you are not convinced why we have attributed a positive sign to the potential difference across the battery and a negative sign to the potential differences across the two resistors, simply start from any point along the circuit and follow the direction of current: you can easily see that current travels from - to +, that is in the direction of an increasing potential (or, as mentioned in Sect. 11.1.3, of a *positive* potential difference) through the battery, and from + to -, that is in the direction of a decreasing potential (or of a *negative* potential difference), through the resistors.

If you're wondering, I haven't told you about these two laws for a form of sadism, but because we will need them to understand how the *Wheatstone bridge*, an essential component of invasive pressure transducers, works (see Sect. 11.2.2).

Most of what we have discussed so far is about direct current (DC) circuits, i.e., circuits in which current "flows" always in the same direction. We will not deal in this chapter (nor in this book) with alternating current (AC), i.e., the one that arrives to our homes, which periodically inverts its direction and changes its magnitude according to a sinusoidal waveform.

11.2.1 Electrical Resistance and Platelet Aggregometry

An electrical resistance measurement is the working principle of the *multiple electrode aggregometry* (MEA) assay performed by the Multiplate® analyzer (Roche Diagnostics, Basel, Switzerland), one of the point-of-care devices for platelet function testing which have been extensively investigated, and are currently used, in several clinical settings such as trauma and cardiac surgery [2–5]. This technology is also referred to as whole-blood *impedance* aggregometry. To be precise, electrical impedance (which, like resistance, is measured in Ω) is a term referred to AC circuits, where it indicates something similar to resistance. However, in DC circuits, impedance and resistance would be exactly the same thing (but "resistance" is the commonly used term).

In practice, MEA measures the changes in electrical resistance over a time interval of 6 min between two pairs of electrodes which are submerged in a sample of diluted whole-blood to which a platelet activator has been added and provides a result in terms of area under the curve (AUC) of resistance by time (see Fig. 11.4). In other words, remembering the few notions of calculus mentioned in Chap. 1, it could be said that the AUC (expressed in arbitrary units U) is the definite integral from 0 to 6 (minutes, that is the duration of the test) of the curve which describes electrical resistance R (expressed in arbitrary units AU) as a function of time t (in minutes):

$$\text{AUC} = \int_0^6 R(t)\,\mathrm{d}t \qquad (11.14)$$

Different tests can be performed according to the specific activator used: *ASPI test* with arachidonic acid (AA), *ADP test* with adenosine diphosphate (ADP), *TRAP* test with thrombin receptor-activating peptide (TRAP), and others. As activated platelets adhere to and aggregate on the electrodes, resistance increases (Fig. 11.4a). However, resistance will increase to a much lesser extent (and, accordingly, the AUC will be much lower) if the patient is taking an antiplatelet medication which acts on that specific pathway, or in case of platelet disorders (Fig. 11.4b). In fact, pharmacologically "inhibited," or anyhow dysfunctional, platelets will adhere much less to the electrodes (and will aggregate less with each other). For example,

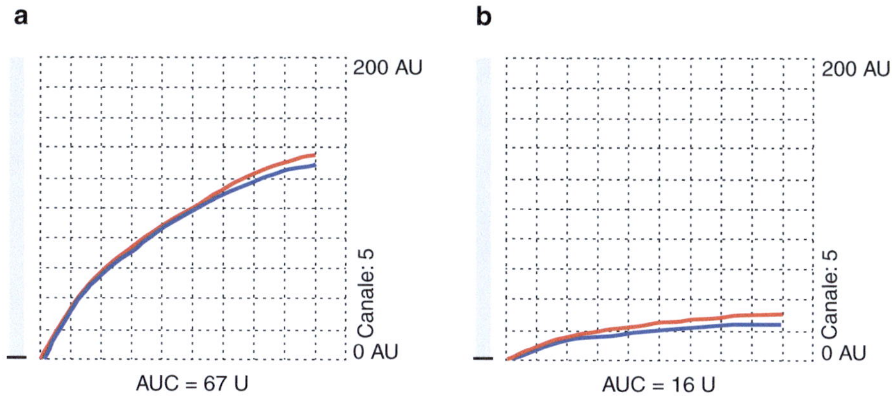

Fig. 11.4 Multiplate® ADP test (Roche Diagnostics, Basel, Switzerland) in two patients on clopidogrel therapy. (**a**) Non-responder (a similar result could be found in a patient who has recently stopped, or who don't take, clopidogrel or other $P2Y_{12}$ receptor antagonists); (**b**) Responder. This patient is also at high risk of postoperative bleeding in case of immediate cardiac surgery (AUC below 31 U). The curves indicate the increase in electrical resistance (in arbitrary units AU) over time as activated platelets adhere to and aggregate on the electrodes. Blue and red curves refer to the two different pairs of electrodes in the same blood sample and must roughly coincide for the test to be valid (internal control). *ADP* adenosine diphosphate; *AUC* area under the curve

the ADP test allows to identify patients who are "non-responders" to antiplatelet drugs which act via the ADP receptors (i.e., the $P2Y_{12}$ receptor antagonists clopidogrel, ticagrelor, and prasugrel); in these patients, the AUC for the ADP test will be above a "therapeutic" cutoff despite they take the usual doses of these drugs. As another example, the ADP test can be used to predict the risk of postoperative bleeding in patients on preoperative $P2Y_{12}$ receptor antagonist therapy undergoing cardiac surgery; in particular, patients with an AUC below 31 U before surgery have been shown to be at higher risk of postoperative bleeding and platelet transfusion [6]. Recently, a platelet aggregometry module (ROTEM® Platelet) which works on the same "impedance" principle (but with different names for the specific tests and different output units) became available in combination with the ROTEM delta® thromboelastometry device (TEM International, Munich, Germany) [7].

11.2.2 Electrical Resistance and Invasive Pressure Monitoring: The Wheatstone Bridge

As mentioned in Chap. 9, pressure transducers that we commonly use for invasive pressure monitoring in the intensive care unit or in the operating room consist of a membrane through which an electric current runs; as pressure and its changes (which are transmitted unaltered along the fluid-filled system according to the Pascal's principle) distort the transducer's membrane, its electrical resistance

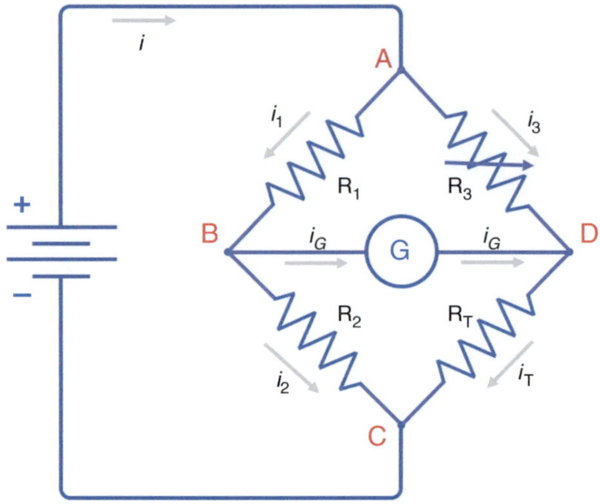

Fig. 11.5 A Wheatstone bridge (see text). i_1 and i_2 are the currents which run through the two fixed resistors with resistance R_1 and R_2, respectively; i_3 is the current through the adjustable resistor (with variable resistance R_3); i_T is the current through the pressure transducer membrane (which has a resistance R_T that varies according to the membrane distortion due to pressure); i_G is the current measured by a galvanometer (G). The bridge is said to be balanced if $i_G = 0$

undergoes very small changes which can be measured, thanks to a particular electric circuit called *Wheatstone bridge* (Fig. 11.5) [8–10]. These changes in electrical resistance are then electronically converted into pressure values. Now that we have discussed electric current, potential, resistance, and learned about electric circuits, resistors, and Kirchhoff's laws, we are ready to understand in detail how a Wheatstone bridge works.

As shown in Fig. 11.5, a Wheatstone bridge is a circuit consisting of four resistors, two with fixed resistances R_1 and R_2, respectively; one with adjustable resistance R_3 (the oblique arrow on the resistor symbol indicates a *variable*, or *adjustable*, *resistor*), and one with the unknown resistance to be measured, namely the resistance R_T of the transducer membrane in the case of pressure transducers; furthermore, a battery provides a potential difference (and, accordingly, a current along the circuit) and a *galvanometer* measures the current i_G between the two sides (or "legs") of the bridge [10].

The bridge is said to be *balanced* when there is no potential difference between its legs and, accordingly, the current i_G measured by the galvanometer between the junctions B and D is zero. The variable resistance R_3 is continuously adjusted to keep the circuit balanced and, since a galvanometer measures "zero current" with very high accuracy, this allows an accurate measurement of R_T according to the (known) values of R_1 and R_2 and to the value of R_3 which is needed in order to maintain the bridge balanced (as the membrane resistance varies). Let's see how R_T is calculated (always keep an eye on Fig. 11.5).

According to Kirchhoff's current law (look at Fig. 11.3a), it is:

$$i_3 + i_G = i_T \tag{11.15}$$

and

$$i_1 = i_G + i_2 \tag{11.16}$$

where i_1, i_2, and i_3 are the currents which run through the resistors R_1, R_2, and R_3, respectively, and i_T is the current through the pressure transducer membrane.

According to Kirchhoff's voltage law, the algebraic sum of the potential differences across all the elements in each of the two loops ADBA and BDCB must be zero. Remembering that the potential difference across any circuit element is the product of the element's resistance and the current through it (Eq. (11.10)), and considering the direction of current in each element, it is:

$$i_3 R_3 - i_G R_G - i_1 R_1 = 0 \tag{11.17}$$

for the loop ADBA, and

$$i_T R_T - i_2 R_2 + i_G R_G = 0 \tag{11.18}$$

for the loop BDCB (in both equations, R_G is, evidently, the galvanometer resistance).

As stated above, if the bridge is balanced, it is $i_G = 0$. Accordingly, Eqs. (11.15–11.18) become, respectively:

$$i_3 = i_T \tag{11.19}$$

$$i_1 = i_2 \tag{11.20}$$

$$i_3 R_3 - i_1 R_1 = 0 \quad \text{or} \quad i_3 R_3 = i_1 R_1 \tag{11.21}$$

$$i_T R_T - i_2 R_2 = 0 \quad \text{or} \quad i_T R_T = i_2 R_2 \tag{11.22}$$

Rearranging Eqs. (11.21) and (11.22), it is:

$$i_3 = \frac{i_1 R_1}{R_3} \tag{11.23}$$

and

$$i_T = \frac{i_2 R_2}{R_T} \tag{11.24}$$

Finally, for the bridge to be balanced it must be $i_3 = i_T$ (Eq. 11.19), that is:

$$\frac{i_2 R_2}{R_T} = \frac{i_1 R_1}{R_3} \tag{11.25}$$

Remembering that it must also be $i_1 = i_2$ (Eq. 11.20) and substituting i_1 with i_2 (or vice versa) so that the two terms cancel out, Eq. (11.25) becomes:

$$\frac{R_2}{R_T} = \frac{R_1}{R_3} \tag{11.26}$$

Resolving for R_T:

$$R_T = \frac{R_2 R_3}{R_1} \tag{11.27}$$

And here's how the resistance of the transducer membrane is calculated according to the values R_1, R_2, and R_3.

11.2.3 Electrical Resistance and Temperature Measurement: Thermistors

The resistivity ρ of any material and, accordingly, the resistance R of any object which is made of that material (see above), varies roughly linearly with temperature. However, these are very small variations. But there are materials such as metal oxides whose resistivity varies (it can either decrease or increase) significantly and nonlinearly (exponentially) with temperature: these materials are used to make "temperature sensitive resistors," called *thermistors* (Fig. 11.6) [9, 11, 12], which have a lot of industrial and technological applications (one of the most intuitive is, after calibration, as a thermometer in electric/electronic circuits). All anesthesiologists and intensivists who have used a pulmonary artery catheter (PAC) at least once know that there is a thermistor at its tip (located in the pulmonary artery) which measures the changes in blood temperature over time after injection of a fluid bolus in the right atrium in order to estimate cardiac output according to the so-called modified *Stewart-Hamilton equation* (we discussed in Chap. 10 the laws of thermodynamics which are behind this *thermodilution* method). Most temperature probes (e.g., esophageal or nasopharyngeal probes) we currently use in clinical practice are also based on thermistors [12, 13].

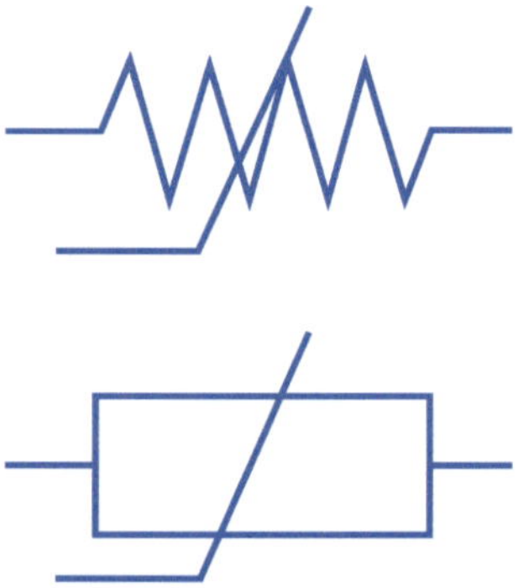

Fig. 11.6 Two different ways to indicate a thermistor in a circuit diagram. The notation $+t°$ or $-t°$ can be added to both symbols to indicate a thermistor whose resistance increases or decreases with temperature, respectively

11.3 Other Temperature Probes: Thermoelectric Effect and Thermocouples

The significant change in the electrical resistance of thermistors with temperature is not the only circumstance characterized by a link between electricity and thermal phenomena. Another one, which is well known to everybody, is the production of *heat* whenever an electric current runs through a conductor (the so-called *Joule effect* or *Joule heating*); this phenomenon gave to mankind the incandescent light bulbs we used for decades (the "trick" is that there is no oxygen in the bulb so the filament heats up and generates light but does not burn). However, the term *thermoelectric effect* is reserved for a set of three different (but interconnected) phenomena concerning some relationship between heat and electricity. One of them, the so-called *Seebeck effect*, is the working principle of *thermocouples*, other circuit elements which, like thermistors, are commonly used in clinical practice as temperature probes [12, 13].

Without going into too much detail, the Seebeck effect can be described as follows: if you build a ring made up of two different metal conductors like the one in Fig. 11.7, a (very small) electric current will develop along it (although there is no source of potential difference, such as a battery, in the circuit) if the two junctions between the different metals are kept at different temperatures, while there will be no current if all parts of the ring have the same temperature. If you cut the ring at any point and connect the free ends to a *voltmeter* (i.e., an instrument that measures potential difference), you will detect a potential difference which is proportional to the temperature difference between the two junctions. Based on this principle, a thermocouple allows the measurement of temperature at one junction between two metals (relative to a reference temperature at the junctions with a voltmeter) according to the potential difference detected (Fig. 11.8); in fact, the relationship between the temperature difference and the voltage generated only depends on the two metals chosen. Thermocouples have different applications, e.g., as temperature meters in home thermostats. Those used for temperature monitoring in clinical practice (e.g., esophageal or tympanic probes) are usually made of copper and a copper-nickel alloy called constantan (type T thermocouple); this type of thermocouple is very cheap and suitable for measurement in the range of body temperatures, but it is limited to single-patient use since constantan is a very fragile material [12].

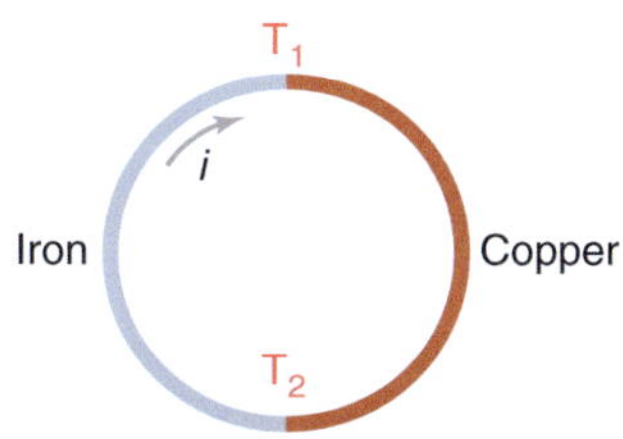

Fig. 11.7 Seebeck effect: in a ring made up of two different metals, a current i can be detected whenever the temperatures T_1 and T_2 at the two junctions between the two metals are different

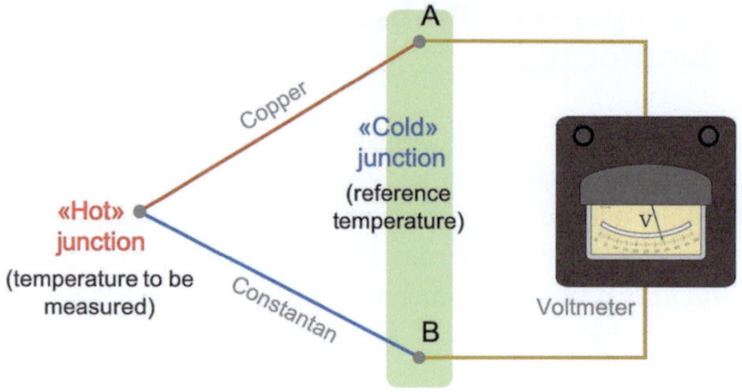

Fig. 11.8 Schematic representation of a type T thermocouple (see text). According to the Seebeck effect, temperature at any point in which the "hot junction" is located (e.g., the esophagus) can be measured (as the difference with that at the "cold junction") according to the potential difference measured by a voltmeter between the points A and B

References

1. Feynman RP, Leighton RB, Sands M. The Feynman lectures on physics: the new millennium edition. New York: Basic Books; 2011. p. 4–7.
2. Casso G, Lanzi F, Marcucci CE. Point-of-care platelet function tests. In: Marcucci CE, Schoettker P, editors. Perioperative hemostasis. Cham, Switzerland: Springer; 2015. p. 45–63.
3. Hanke AA, Horstmann H, Wilhelmi M. Point-of-care monitoring for the management of trauma-induced bleeding. Curr Opin Anaesthesiol. 2017;30(2):250–6.
4. Mueller T, Dieplinger B, Poelz W, et al. Utility of whole blood impedance aggregometry for the assessment of clopidogrel action using the novel multiplate analyzer-comparison with two flow cytometric methods. Thromb Res. 2007;121(2):249–58.
5. Ranucci M, Pistuddi V, Di Dedda U, et al. Platelet function after cardiac surgery and its association with severe postoperative bleeding: the PLATFORM study. Platelets. 2019;30(7):908–14.
6. Ranucci M, Baryshnikova E, Soro G, et al. Multiple electrode whole-blood aggregometry and bleeding in cardiac surgery patients receiving thienopyridines. Ann Thorac Surg. 2011;91(1):123–9.
7. Ranucci M, Baryshnikova E, Crapelli GB, et al. Electric impedance platelet aggregometry in cardiac surgery patients: a comparative study of two technologies. Platelets. 2016;27(3):185–90.
8. Magder S. Invasive intravascular hemodynamic monitoring: technical issues. Crit Care Clin. 2007;23:401–14.
9. Szocik JF, Barker SJ, Tremper KK. Fundamental principles of monitoring instrumentation. In: Miller RD, editor. Miller's anesthesia. 6th ed. Philadelphia, PA: Elsevier Churchill Livingstone; 2005. p. 1191–225.
10. Aston D, Rivers A, Dharmadasa A. Equipment in anaesthesia and critical care. Banbury: Scion Publishing Limited; 2014. p. 375.
11. Aston D, Rivers A, Dharmadasa A. Equipment in anaesthesia and critical care. Banbury: Scion Publishing Limited; 2014. p. 205.
12. Sessler DI. Perioperative temperature monitoring. Anesthesiology. 2021;134(1):111–8.
13. Sessler DI. Perioperative thermoregulation and heat balance. Lancet. 2016;387(10038): 2655–64.

Spark plugs, Computer Keyboards, and Defibrillators: Capacitors

12

Contents

Where's Physics

Daily life	Spark plugs (for gasoline cars or scooters) Camera flashes Computer keyboards Touchscreen smartphones
Physics involved	Electric field, Electric potential energy, Current, Circuits (see Chap. 11) Capacitance Capacitors
Clinical practice	Defibrillators

12.1 Storing Electric Energy: Capacitors

(Notice: it is strongly recommended that you have already read Chap. 11 before starting!). Figure 12.1 shows a spark plug like those for gasoline cars or scooters (a), a camera with a big flash (b), a computer keyboard (c), a touchscreen smartphone (d), an automated external defibrillator (AED) (e), and a manual biphasic defibrillator like those found in all intensive care units and operating rooms (f) (I could have shown you only one type of defibrillator, but the figure would have been ugly without a sixth picture!). These objects are very different from each other (with the

A. Pisano, *Physics for Anesthesiologists and Intensivists*,
https://doi.org/10.1007/978-3-030-72047-6_12

Fig. 12.1 All these devices contain (and work thanks to) one or more capacitors (**a–f**: see text)

exception of the two defibrillators, of course). However, they have at least one thing in common: they all contain (and work thanks to) a *capacitor*.

A capacitor is a circuit element used to store *energy* in the form of an *electric field* (see Chap. 11) and consists, in general, of two isolated conductors (called plates) facing each other and separated by a distance *d* (see Fig. 12.2). In particular, Fig. 12.2 shows a *parallel-plate capacitor* connected to a battery (i.e., as mentioned in Chap. 11, a device which maintains a potential difference between its terminals); when a capacitor is placed into an electric circuit like this, electrons start to flow

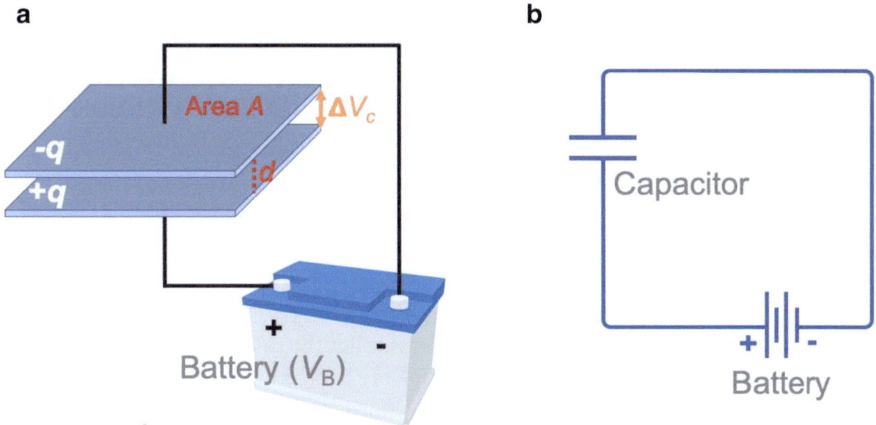

Fig. 12.2 A parallel-plate capacitor connected by wires to a battery (**a**) and related circuit diagram (**b**). ΔV_C Potential difference between the two plates of the capacitor; V_B Voltage (potential difference) of the battery; $\pm q$ Charges of the two plates of the capacitor

from one plate of the capacitor to the positive (+) terminal of the battery (i.e., that with higher electric potential, see Sect. 11.1.3) and from the negative (−) terminal of the battery (i.e., that with lower potential) to the other plate of the capacitor. Accordingly, the first plate becomes more and more positively charged, while the second one becomes more and more negatively charged. At the same time, the potential difference ΔV_C between the two plates of the capacitor progressively *increases*; when ΔV_C equals the potential difference V_B of the battery, the positively charged plate will be at the same potential of the positive terminal of the battery, while the negatively charged plate will be at the same potential of the negative terminal of the battery; hence, current stops and the capacitor is said to be *fully charged* (with the two plates having charges of the same magnitude but opposite signs, i.e., $+q$ and $-q$, respectively).

As it can be guessed, the potential difference ΔV_C between the two plates of the capacitor is directly proportional to the charge q of the capacitor:

$$q = C\ \Delta V_C \tag{12.1}$$

The proportionality constant C is called the *capacitance* of the capacitor. According to Eq. (12.1), its unit in the International System of Units (SI) is the coulomb (i.e., the SI unit of charge) per volt (i.e., the SI unit of potential difference), which is called *farad* (F). In practice, capacitance tells us how much charge must be put on the plates of a capacitor to produce a certain potential difference between them.

For a parallel-plate capacitor, capacitance is directly proportional to the plate area A and inversely proportional to the plate separation d (see Fig. 12.2):

$$C = \varepsilon_0 \frac{A}{d} \tag{12.2}$$

where ε_0 is the permittivity constant that we have already encountered in Chap. 11 (Sect. 11.1.1). In practice, the smaller the distance d between the two plates of a capacitor, the greater the charge q on the plates for a given potential difference ΔV_C across them (just have another look to Eqs. (12.1) and (12.2)). This translates into the notion that the electric field strength E inside a capacitor is inversely proportional to the distance d between its plates:

$$E = \frac{\Delta V_c}{d} \tag{12.3}$$

That is, for any potential difference ΔV_C across the two plates of a capacitor, the capacitor's electric field will be stronger the smaller the distance between its plates.

The (electric potential) energy (see Chap. 11, Sect. 11.1.3) stored in a capacitor can be calculated according to the *work* that must be done (in our case by the battery) to transfer the charge q from one plate of the capacitor to the other one *against* the electric field which is gradually created inside the capacitor. Imagine that a capacitor has a certain charge q_0 on its plates and, accordingly, a certain potential difference ΔV_0 between them. According to Eq. (11.7) (i.e., the definition of potential difference discussed in Chap. 11), the work W_n needed to transfer an additional small charge q_n from one plate to the other (i.e., to make this charge move from a lower to a higher potential) is:

$$W_n = q_n \Delta V_0 \tag{12.4}$$

Of course, this transfer will increase the potential difference between the two plates by a certain amount. Accordingly, the work needed to transfer another small additional charge will be greater. In practice, starting from a discharged capacitor, the work needed to transfer charges from one plate to the other progressively increases (as the electric field against which the charges move becomes more and more intense). Using integral calculus (or even a simple geometric reasoning), it is easy to show that the total work W needed to charge the capacitor with a charge q (and to a potential difference ΔV_C) is:

$$W = \frac{1}{2} q \Delta V_C \tag{12.5}$$

which, according to Eq. (11.6) (Chap. 11), is also the electric potential energy U stored in it:

$$U = \frac{1}{2} q \Delta V_C \tag{12.6}$$

Replacing q with $C\,\Delta V_C$ (according to Eq. 12.1), Eq. (12.6) becomes:

$$U = \frac{1}{2} C \left(\Delta V_C\right)^2 \tag{12.7}$$

that is, the electric potential energy stored into a capacitor which has been charged to a certain potential difference ΔV_C is directly proportional to its capacitance C and, of course, is greater the greater the potential difference ΔV_C between its plates. At this point, it is also easy to show how this energy is associated with the magnitude of the electric field E of the capacitor. In fact, replacing C with $\varepsilon_0 A/d$ (according to Eq. (12.2)) and ΔV_C with dE (according to Eq. (12.3)), Eq. (12.7) becomes:

$$U = \frac{1}{2}\frac{\varepsilon_0 A}{d} d^2 E^2 = \frac{1}{2}\varepsilon_0 A d E^2 \tag{12.8}$$

Looking at Fig. 12.2a, it is easy to realize that Ad is the volume V of the capacitor. Remembering that in Chap. 11 (Sect. 11.1.2), we defined the *energy density* u_E as the electric potential energy U per unit volume, Eq. (12.8) can be rewritten as follows:

$$\frac{U}{Ad} = \frac{U}{V} = u_E = \frac{1}{2}\varepsilon_0 E^2 \tag{12.9}$$

which is Eq. (11.5) already encountered in Chap. 11.

Summing up, charge separation in the capacitor creates an electric field with an associated electric potential energy which, as we have seen, depends on the capacitor's capacitance and on the potential difference applied.

12.2 Computer Keyboards, Smartphones, and Defibrillators

The advantage of capacitors lies in the fact that the energy they store, which is accumulated (relatively slowly) during a certain period of time (and can theoretically remain stored in the capacitor indefinitely), can be released all together in a very short time (all that needs to be done is to short-circuit the capacitor, i.e., to connect its plates between each other by means of a conductor). And here is the presence of a spark plug, a camera flash, and a defibrillator in the same figure (Fig. 12.1): they all need the quick release of a large energy (and use a capacitor for this purpose) in order to ignite the air-fuel mixture in the cylinders, to generate a very short and intense light that serves to illuminate a dark scene for the time of a photo, and to restore the sinus rhythm in the heart [1–3], respectively.

Figure 12.3 shows a very simplified circuit diagram of an *external defibrillator* [1, 2]; with the switch in position A, the power supply charges the capacitor (the figure shows a battery as power supply, but this is typically a *step-up transformer* which provides a potential difference of up to 5000 V, associated to a *rectifier* which converts alternating current [AC] into direct current [DC]); with the switch in position B, the energy stored into the capacitor is discharged to the patient as an electric current (the "DC shock") whose intensity depends on the transthoracic electrical resistance (or, more generally, impedance) according to Eq. (11.10) (Chap. 11) [2, 3]. Many other elements are missing from the figure, including for example an

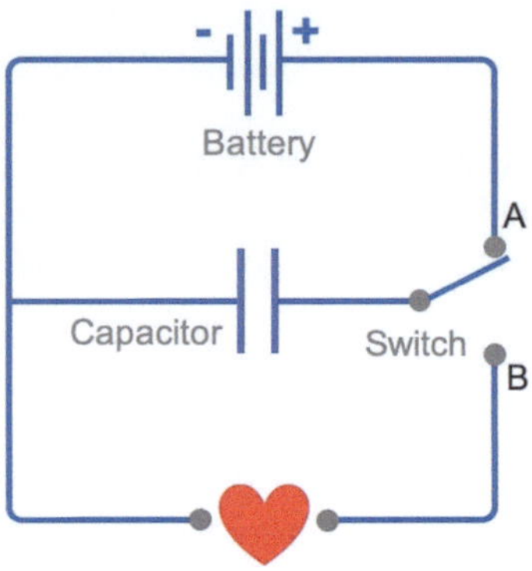

Fig. 12.3 (Very) simplified circuit diagram of an external defibrillator (see text)

inductor which sustains the current produced by the capacitor discharge for several milliseconds (otherwise, it would exponentially decay in a very short time) [2]. However, we leave these details to the engineers.

When we set the "energy" on an external defibrillator (e.g., 200 joules), we're actually setting the potential difference (voltage) to which the capacitor contained in the device is charged: each potential difference applied will correspond to an energy stored in the capacitor (whose capacitance is known) according to Eq. (12.7). Capacitors contained in biphasic external defibrillators usually have capacitances of 100–200 microfarad (μF) [3]. As an example, to charge to 200 J a defibrillator whose capacitor has a capacitance of 100 μF (=0.0001 F) we must select the following potential difference to be applied between the plates of the capacitor:

$$\Delta V_C = \sqrt{\frac{2U}{C}} = \sqrt{\frac{2 \times 200}{0.0001}} = 2000\,V \tag{12.10}$$

(Equation (12.10) is simply Eq. (12.7) resolved for ΔV_C).

We have seen what is the role of the capacitor in a defibrillator (and we can imagine a similar role in a spark plug or in a camera flash). But what about computer keyboards and touchscreen smartphones? There are no high energies to release here. However, also these devices rely on capacitors. Each key on a computer keyboard is associated to a parallel-plate capacitor; when you press a key, the two plates get closer and, according to Eq. (12.2), the capacitor's capacitance increases. Without going into detail, the computer circuit can detect this increase in capacitance, thus recognizing the key which has been pressed. Finally, one type of touchscreen technology, called "capacitive touchscreen," is also based on capacitors (hence the name). As we all know, we humans are good conductors of electricity. Like any conductor, we also have our own capacitance (so far, we have been referring to capacitors, but the concept of capacitance applies to every conductor). Roughly, when we touch the screen of a (capacitive) touchscreen device such as a smartphone, we add our own capacitance to one or more among a set of many small capacitors which are located below the screen (or it is our finger itself acting as the second plate of each capacitor); once again, the circuit can detect this change in capacitance, thus recognizing the point of the screen which has been touched [4].

References

1. Aston D, Rivers A, Dharmadasa A. Equipment in anaesthesia and critical care. Banbury: Scion Publishing Limited; 2014. p. 319.
2. Williams D. Physical principles of defibrillators. Anaesth Intensive Care Med. 2012;13(8):384–7.
3. Weisz MT. Physical principles of defibrillators. Anaesth Intensive Care Med. 2009;10(8):367–9.
4. Kakalios J. The Physics of everyday things. New York: Broadway Books; 2017.

Part V

Forces in Action

Doors, Steering Wheels, Laryngoscopes, and Central Venous Catheters: The Moment of a Force

13

Contents

Where's Physics

Daily life	Doors and doorknobs Steering wheels Pliers, nutcrackers, and tweezers
Physics involved	Vectors vs. scalars Components of vectors Moment of a force (torque) and levers
Clinical practice	Inserting a central venous catheter Proper use of the laryngoscope (tracheal intubation and tooth injury)

13.1 Vectors, Vector Sum, and Components of a Force

Physical quantities such as *force*, but also displacement, velocity, acceleration (and, as discussed in Chap. 11, electric field), to name but a few, cannot be defined by a single value: in addition to their "magnitude", in fact, we need to specify their *direction*. Quantities of this kind are called *vectors* (and are usually indicated by an arrow over their symbol). Conversely, quantities which are completely described by a number (accompanied by a unit and, sometimes, by a sign) are called *scalars*: mass,

A. Pisano, *Physics for Anesthesiologists and Intensivists*,
https://doi.org/10.1007/978-3-030-72047-6_13

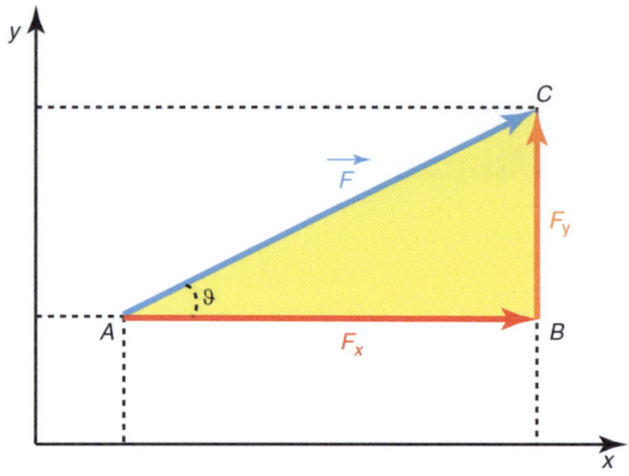

Fig. 13.1 Components of a force *F* along the *x* and *y* axes of a rectangular coordinate system

energy, volume, density, temperature, charge and, as we saw in Chap. 11, electric potential, are some examples.

The *projection* of a vector (e.g. a force) along an axis is called the *component* of the vector on that axis. To find the component of a vector (or, as it is said, to "resolve" the vector) we can place it on a rectangular (Cartesian) coordinate system and use *trigonometry* (see Chap. 1).

Figure 13.1 shows the components of a force $\vec{F}$ along the *x* and *y* axes. Called ϑ the angle made by the force vector with the *x* axis, and remembering the definitions of *sine* and *cosine* (see Box 13.1 or also Chap. 1), is:

$$F_x = F\cos\vartheta \tag{13.1}$$

and

$$F_y = F\sin\vartheta \tag{13.2}$$

where F_x and F_y are the *x* component and the *y* component of $\vec{F}$, respectively.

For example, in Chap. 6 (Box 6.1) we defined the work *W* (done by a force $\vec{F}$ that causes a displacement) as the product of the displacement *d* and the magnitude of the component of the force along the displacement itself (F_d). According to Eq. (13.1), such a component is:

$$F_d = F\cos\vartheta \tag{13.3}$$

where here ϑ is, evidently, the angle between the directions of the force and the displacement. The work is, therefore:

$$W = dF_d = dF\cos\vartheta \tag{13.4}$$

that is, as mentioned in Box 6.1, the *scalar* (or *dot*) *product* of force and displacement (in general, the scalar product of two vectors is the product of the magnitudes of the two vectors and of the cosine of the angle between their directions, or the product of the magnitude of one of the two vectors and the component of the second vector along the direction of the first one).

Unlike scalars, two vectors cannot be summed algebraically. If we imagine that F_x and F_y (Fig. 13.1) are two different forces which acts, at the same time, on an object (e.g. both applied to the point A), their *vector sum* (or *resultant*) can be obtained using the so-called *parallelogram method*: the two vectors are drawn head to tail (as in Fig. 13.1), and the resultant vector is that which goes from the tail of the one to the head of the other. It is easy to see that, in this case, the resultant vector is $\vec{F}$. In this particular case, in which the two forces are perpendicular to each other, the magnitude of their vector sum can be calculated by means of the *Pythagorean theorem* (see Chap. 1).

Box 13.1: Remembrances of Trigonometry (to Explain Eqs. 13.1 and 13.2)

As discussed in Chap. 1, *sine* and *cosine* are two trigonometric functions of an angle which are classically defined in the context of a right triangle. In particular, the sine of an angle ϑ is the ratio between the length of the side of the triangle which is opposite to ϑ and that of the hypotenuse, while the cosine of ϑ is the ratio between the length of the side which is adjacent to ϑ and that of the hypotenuse. Hence, in Fig. 13.1, is:

$$\sin\vartheta = BC / AC \text{ and } \cos\vartheta = AB / AC$$

or

$$BC = AC\sin\vartheta \text{ and } AB = AC\cos\vartheta$$

which are none other than Eqs. (13.2) and (13.1), respectively (in fact, in the figure it is $BC=F_y$, $AB=F_x$, and $AC=F$).

13.2 Pliers, Nutcrackers, Tweezers (and so on): Moment of a Force and the Levers

Why can we unscrew a bolt with the help of a wrench but not with bare hands? After all, which help do we get from an inanimate tool in terms of strength? Have you ever tried to open a heavy door by pushing it from a point very close to its hinges? It takes a lot more force than if you use the doorknob (it is not a coincidence, indeed, that doorknobs are where they are). Have you ever wondered why old trucks or buses had much larger steering wheels than cars? And why truckers are no longer the strong and muscled men than once?

Consider a body, such as a metal bar, that can rotate about an axis A (Fig. 13.2). A force $\vec{F}$ which is applied at a point P will cause rotation of such a body according to its direction and magnitude, but also depending on the position of its point of

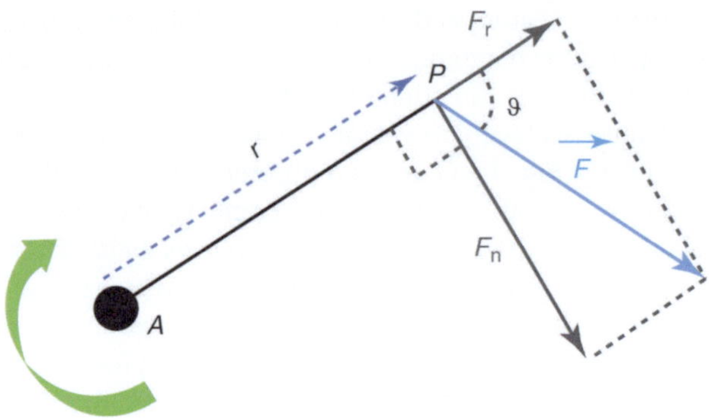

Fig. 13.2 A body such as a metal bar rotates about an axis A under the effect of a force F applied at a point P (which is located at a distance r from A). F_r radial component of F; F_n tangential (or normal) component of F

application P with respect to the rotation axis A, which is defined by the position vector r. As shown in Fig. 13.2, the force $\vec{F}$ can be resolved into two components: the *radial* component F_r, pointing along r, and the *tangential* (or *normal*) component F_n, perpendicular to r. Evidently, only the latter will rotate the metal bar. According to what discussed above, the magnitude of this normal component is:

$$F_n = F \sin \vartheta \tag{13.5}$$

where ϑ is the angle that the directions of r and F make with each other.

In order to consider all factors involved in the rotation caused by $\vec{F}$ (magnitude, direction and distance from the rotation axis), we define a vector called *moment* $\overrightarrow{M}$ (or *torque*) as the product of the distance r and the normal component $\overrightarrow{F_n}$:

$$\overrightarrow{M} = r\overrightarrow{F_n} \tag{13.6}$$

Since, according to Eq. (13.5), the magnitude of $\overrightarrow{F_n}$ is $F \sin \vartheta$, the magnitude of $\overrightarrow{M}$ will be:

$$M = r F \sin \vartheta \tag{13.7}$$

Therefore, the "actual" force causing rotation (i.e. the moment or torque of F) is as much more magnified the greater is the distance from the rotation axis.

This is the reason why the steering wheel of the old trucks or buses (who had no power steering) was much larger: a greater distance from the rotation axis allowed a greater "effective" force (i.e. a greater moment) with the same effort by the driver. With the introduction of power steering, however, all this is no longer necessary (and truckers can even be weakling!). Moreover, the kind of help provided by a wrench and the advantage of having doorknobs away from the hinge line are certainly clear now.

This is also why Archimedes said that it would raise the World with a *lever* long enough. On the principle of torque, in fact, is based the functioning of *levers* (Fig. 13.3), which are classified according to the position of the fulcrum (i.e. the rotation axis) relative to the point of application of a force F which acts in order to overcome a "resistance" R, and that of R itself (for practical reasons, the arrow above the symbols F and R is omitted from here onwards: we know that they are both vectors). In a *first class lever* such as a pliers (Fig. 13.3a), the fulcrum is located

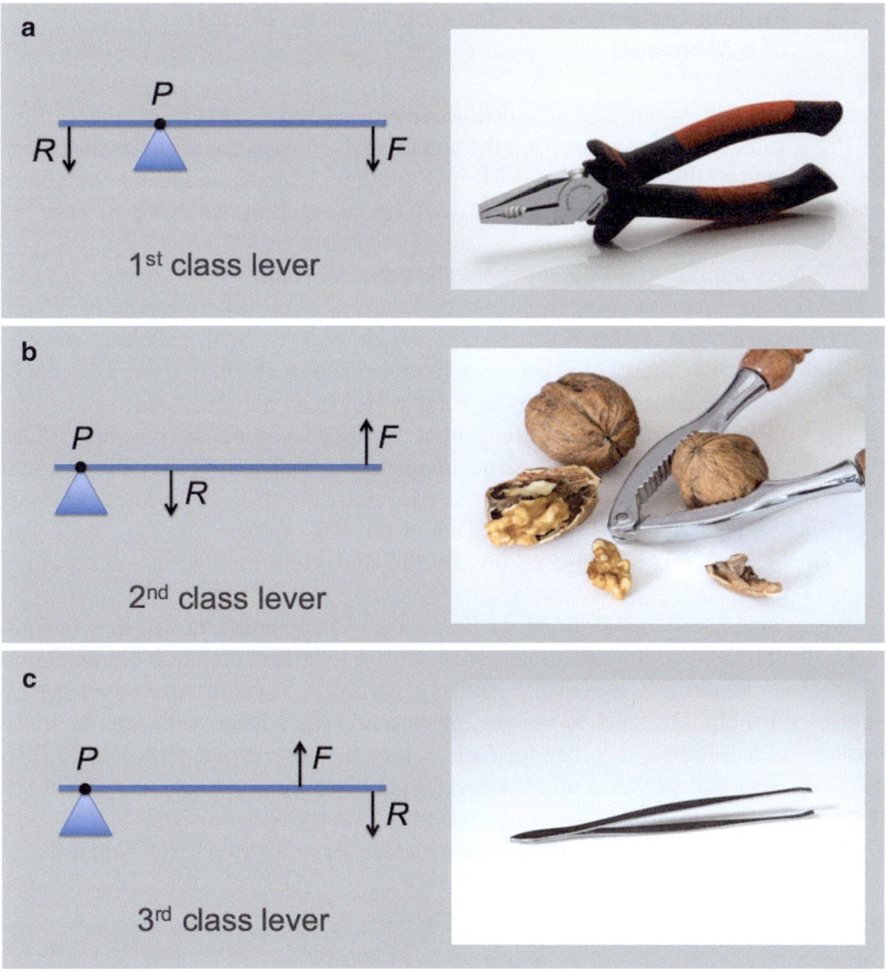

Fig. 13.3 Schematic representation and examples of the three basic types of levers: (**a**) first class lever; (**b**) second class lever; (**c**) third class lever. *P* fulcrum; *F* force; *R* resistance

between the force and the resistance. This kind of lever is (mechanically) advantageous only if the fulcrum is closer to *R* than to *F* (and, accordingly, the moment of *F* is higher than that of *R*). In a *second class lever* (e.g. a nutcracker, Fig. 13.3b), the resistance is in the middle: this lever is always advantageous as *F* is always more distant from the fulcrum as compared to *R*. Finally, a *third class lever* (such as the tweezers shown in Fig. 13.3c) is always disadvantageous since *F* is in the middle (closer to the fulcrum than *R*). Note that the terms "advantageous" and "disadvantageous" refer to the fact that the force applied is, respectively, "magnified" or "weakened" (as compared to the resistance to be overcome), not to the usefulness of the tool: in the case of the tweezers, for example, the "advantage" (of the disadvantage!) lies in a greater delicacy and precision.

13.3 Bend a Guidewire or Blow up a Tooth: Matter of a Moment!

We all know that, when placing a *central venous catheter*, the advancement of the dilator is a delicate step. Apart from the potentially serious complications which can be associated to this maneuver [1–3], it is common experience that, if the dilator (and, consequently, the guidewire inside) bends, at best we have to start all over again.

The force applied to the dilator should be perfectly parallel to it (so that the angle ϑ between the direction of the force and the long axis of the dilator is equal to 0 and, accordingly, the moment is null): in this way, in fact, we will not cause rotation about the insertion site (which acts as a fulcrum) and, accordingly, the dilator will be not subjected to a torque tending to bend it. However, our thrust may not be perfectly parallel, so that a small tangential component appears (see Fig. 13.4). Although small, this component (perpendicular to the long axis of the dilator) could be enough to bend the dilator if its distance from the insertion site is large, since its moment will be high (Fig. 13.4a). Here's how physics explains why the proper way to hold the dilator is as close as possible to the insertion site, as we all know (Fig. 13.4b).

We all also know that *tooth injury* (including either enamel, crown or root fractures, dental subluxation, and tooth avulsion) is a relatively common complication of *tracheal intubation*, especially with classic direct laryngoscopy [4–7]. Many authors warn clinicians not to use the *laryngoscope* as a lever, with the maxillary incisors as a fulcrum, in order to avoid injury to these dental elements [7–13]. However, now that we know all (or almost) about levers and the moment of a force, it should be clear that the major concern is not to use the upper incisors as a fulcrum [7–10] (and, accordingly, the laryngoscope as a first class lever), but rather, doing

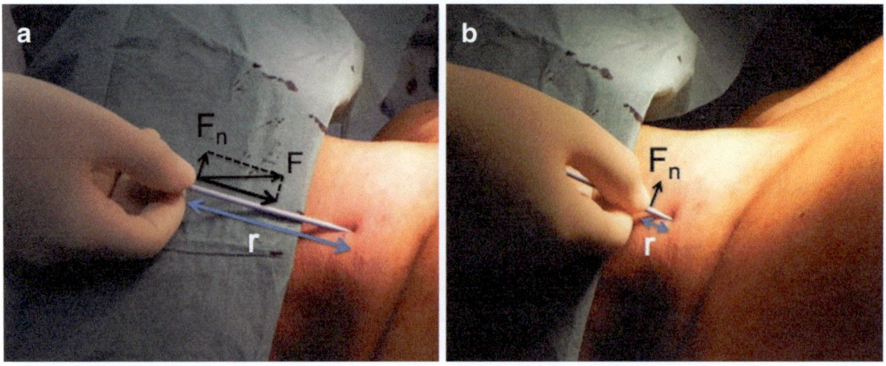

Fig. 13.4 Insertion of a central venous catheter. (**a**) Gripping the dilator distally, it acts as a second class lever; therefore, the probability of winning its resistance (and bend it and the guidewire) increases. (**b**) Gripped more proximally to the insertion point, the dilator is more similar to a third class lever or, in any case, the moment of the force *F* we apply is much less and, consequently, the risk of bending the dilator is much lower. F_n normal component of *F*

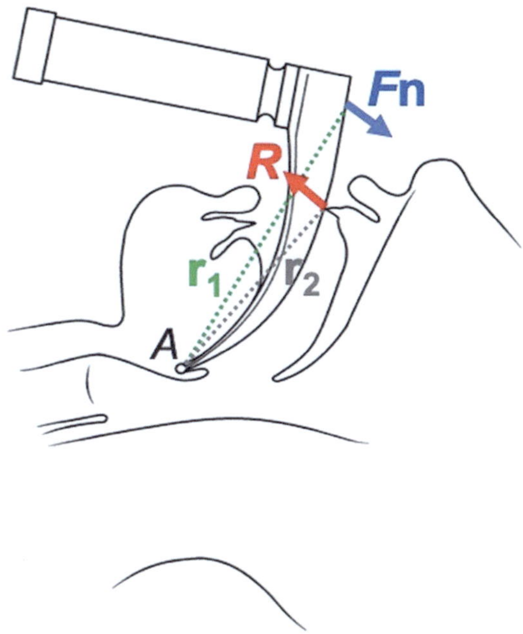

Fig. 13.5 When using mistakenly the laryngoscope as a lever by leaning on maxillary incisors, it may dangerously become a second class lever with fulcrum A at the level of the epiglottic vallecula (see text): in fact, since it is $r_1 > r_2$, the moment of the normal component F_n of the force that we apply (which has magnitude $M_F = r_1F_n$) will be greater than the moment of an equal and opposite resistance R from the teeth (with magnitude $M_R = r_2R$). This increases the risk of tooth injury. The laryngoscopy illustration is courtesy of Rasul Ashurov, MD, cardiac surgery resident, University of Campania "Luigi Vanvitelli", Naples, Italy

so, to transform the laryngoscope *in a second class lever*, with the *epiglottic vallecula* as a fulcrum, so as to magnify the force acting against the "resistance" opposed by the teeth. It is clear from Fig. 13.5, in fact, that the situation that arises when the epiglottis does not rise further and we continue to make strength is very similar to that of the nutcracker of Fig. 13.3b (with the difference that we do want to break the walnut, but preferably not the teeth!). This was just to be accurate from a physical standpoint: anyway, stay away from the upper incisors during laryngoscopy!

References

1. Robinson JF, Robinson WA, Cohn A, Garg K, Armstrong JD 2nd. Perforation of the great vessels during central venous line placement. Arch Intern Med. 1995;155(11):1225–8.
2. Oropello JM, Leibowitz AB, Manasia A, Del Guidice R, Benjamin E. Dilator-associated complications of central vein catheter insertion: possible mechanisms of injury and suggestions for prevention. J Cardiothorac Vasc Anesth. 1996;10(5):634–7.

3. Bowdle A. Vascular complications of central venous catheter placement: evidence-based methods for prevention and treatment. J Cardiothorac Vasc Anesth. 2014;28(2):358–68.
4. Tan Y, Loganathan N, Thinn KK, Liu EHC, Loh NW. Dental injury in anaesthesia: a tertiary hospital's experience. BMC Anesthesiol. 2018;18(1):108.
5. Newland MC, Ellis SJ, Peters KR, Simonson JA, Durham TM, Ullrich FA, Tinker JH. Dental injury associated with anesthesia: a report of 161,687 anesthetics given over 14 years. J Clin Anesth. 2007;19(5):339–45.
6. Schieren M, Kleinschmidt J, Schmutz A, Loop T, Staat M, Gatzweiler K-H, Wappler F, Defosse J. Comparison of forces acting on maxillary incisors during tracheal intubation with different laryngoscopy techniques: a blinded manikin study. Anaesthesia. 2019;74(12):1563–71.
7. Engoren M, Rochlen LR, Diehl MV, Sherman SS, Jewell E, Golinski M, Begeman P, Cavanaugh JM. Mechanical strain to maxillary incisors during direct laryngoscopy. BMC Anesthesiol. 2017;17(1):151.
8. de Sousa JM, Mourão JI. Tooth injury in anaesthesiology. Braz J Anesthesiol. 2015;65(6):511–8.
9. Fowler RA, Pearl RG. The airway: emergent management for nonanesthesiologists. West J Med. 2002;176(1):45–50.
10. Ghali GE, Meram AT, Garrett BC. Dental injury: anatomy, pathogenesis, and anesthesia considerations and implications. In: Fox III C, Cornett E, Ghali G, editors. Catastrophic perioperative complications and management. Cham: Springer; 2019. p. 83–94.
11. Collins SR. Direct and indirect laryngoscopy: equipment and techniques. Respir Care. 2014;59(6):850–62.
12. Mourão J, Neto J, Luís C, Moreno C, Barbosa J, Carvalho J, Tavares J. Dental injury after conventional direct laryngoscopy: a prospective observational study. Anaesthesia. 2013;68(10):1059–65.
13. Bhalla T, Maa T, Sawardekar A. Airway management. In: Frendl G, Urman RD, editors. Pocket ICU. Philadelphia, PA: Lippincott Williams & Wilkins; 2013. p. 4.1–9.

14 Friction, Trigonometry, and Newton's Laws: All About Trendelenburg Position

Contents

Where's Physics

Daily life	Motion or motionlessness (of everything) Cars and engines Weight Jets, rockets, rifles, and cannons
Physics involved	Newton's laws Gravity (gravitational force) Normal force Friction Inclined plane (ramp)
Clinical practice	Safety in the Trendelenburg position Endotracheal tube cuff and pilot balloon

A. Pisano, *Physics for Anesthesiologists and Intensivists*,
https://doi.org/10.1007/978-3-030-72047-6_14

14.1 Forces and Motion: Newton's Laws

The *Newton's laws*, also known as *laws of dynamics* (that is the branch of classical mechanics which deals with forces and their effect on motion), have been all already mentioned in previous chapters (e.g., Chaps. 3, 8, and 9). The concept of *friction* has been also faced several times (see, in particular, Chaps. 6 and 8). Moreover, in Chap. 13 we discussed how to resolve a force in its components and how to find the sum (or resultant) of different forces acting on a body. It is now time to put all together (and add something more); this will allow us to take an (enjoyable, I hope) journey to discover how physics can help us to make *Trendelenburg position* (see Fig. 14.1) safe.

Let's start from the Newton's laws.

14.1.1 Newton's First Law

Also referred to as *law of inertia*, *Newton's first law* states that if no *net* force acts on a body, such a body remains at rest if it was at rest, while it continues to move *at a constant speed* (i.e., with uniform rectilinear motion) if it was moving. Alternatively, Newton's first law can be enunciated as follows: a body cannot *accelerate* (i.e., its speed cannot change) if no *net* force acts on it. In reality, the first law can be derived (as a special case) from the *Newton's second law*, which establishes

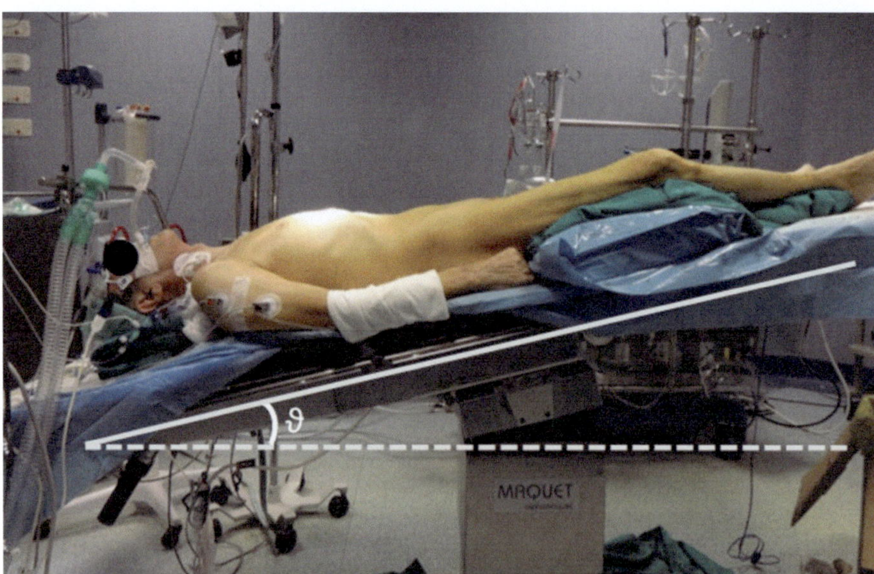

Fig. 14.1 A patient in the Trendelenburg position. The "degree" of tilt refers to the width of the angle ϑ between the operating table and a line parallel to the floor

a quantitative relationship between the force applied to a body and the acceleration caused by this force.

14.1.2 Newton's Second Law

The *net* force acting on a body is the product of its *mass* and *acceleration*:

$$\vec{F} = m\vec{a} \tag{14.1}$$

where the vectors $\vec{F}$ and $\vec{a}$ are, respectively, the net force acting on the body and the acceleration caused by the force itself, and m is the mass of the body.

Equation (14.1) expresses the second law of dynamics (*Newton's second law*), according to which a given force applied to a body, *whatever its speed*, produce an acceleration of such a body (along the same axis of the force) whose magnitude is directly proportional to the force itself. Moreover, Eq. (14.1) allows to define mass as a characteristic of a body which determines "how much" the body accelerates due to a given force applied on it: the higher the mass, the lower the acceleration (see also Chap. 3, Box 3.1).

Evidently, Newton's second law applies regardless of the speed of a body (as stated above) only if we can consider the mass of such a body as *independent from its speed*. According to (Einstein's) *special relativity* (see Chap. 21), the mass of a body *increases* with its speed [1, 2]. However, this becomes evident only for speeds which are not negligibly low as compared to the *speed of light*, that is (in vacuum) about 300,000 km/s (or 186,000 mi/s): at "ordinary" speeds to which we are accustomed, we can peacefully rely on Newton's law.

As mentioned in Chap. 3 (Box 3.1), Eq. (14.1) also allows to define the *weight* W of a body and its relationship with the mass m of the body itself. In fact, if the only force which is applied to the body is its weight (i.e., the magnitude of the *gravitational force,* that is, in our everyday life, the force which pulls everything towards the center of the Earth), then the acceleration produced is the *free-fall acceleration* (namely the acceleration due to *gravity*), and Eq. (14.1) becomes:

$$\vec{W} = m\vec{g} \tag{14.2}$$

where g is the magnitude of the free-fall acceleration that is equal to 9.8 m/s^2 or 32 ft/s^2 (as we have already seen in Chap. 3).

Just to be precise (as mentioned in Chap. 3), it is nowadays generally accepted that, according to *general relativity* (the wonderful "theory" by Albert Einstein, which I suggest to those who are intrigued to read from the author's own words [3], although it could be easier to rely on some clear and enjoyable divulgative books [1, 2, 4, 5]), gravity is not properly a force acting at a distance between two bodies, but rather the result of a *curvature of spacetime* due to the presence of such bodies. However, as discussed in Box 3.1, the (Einstein's) *principle of equivalence* allows us to speak about an "acceleration caused by gravity" (and, hence, to treat gravity as a force) since the acceleration which one experiences due to the gravitational attraction by a massive object is indistinguishable from an acceleration of the same magnitude which is produced by an appropriate force [1]. Anyway, the knowledge of Eq. (14.2) (and simply refer to gravity as an attractive force exerted by the Earth) is all we need for our purposes; accordingly, feel free to ignore the last two sentences.

14.1.3 Newton's Third Law

In its simplest formulation (the so-called *law of action and reaction*), the third law of dynamics (or *Newton's third law*) states that if a body *A* exerts a force (which we can call "action") on a body *B*, then the body *B* exerts, on the body A, a force ("reaction") that is equal in magnitude and opposite in direction to that exerted by *A* on *B*. The two bodies are said to *interact* with one another, while the opposite forces due to such interaction are often referred to as an *action-reaction force pair* or (*Newton's*) *third-law force pair*. When you push with your hand on a wall, you are exerting a force against it, but also the wall is exerting, at the same time, an (equal and opposite) force against you; in fact, if you are wearing roller skates, or if the floor is wet, you could slip in the opposite direction to the wall (and it was the wall that pushed you!). Other common examples of action-reaction force pair are a bouncing ball, the propulsion of jets and rockets, and the recoil of firearms. When you shoot with a rifle, the rifle exerts a force on the bullet, but also the bullet exerts a force on the rifle. As a consequence, *both* objects move (in the opposite direction, since the two forces have opposite directions). One might wonder why the bullet moves much faster (and, accordingly, travels much farther) than the rifle despite the forces that the two objects exert on one another are equal in magnitude; this is due to the much lower mass of the bullet. In fact, according to the so-called *law of conservation of linear momentum* (another aspect of Newton's third law), the total linear momentum (i.e., the product of mass and speed) of a closed and isolated system remains constant. Accordingly, the linear momentum of the rifle must be equal to that of the bullet; hence, the lower the mass, the higher the speed.

We have already mentioned Newton's third law in Chaps. 2 and 9, when discussing about the endotracheal tube cuff. With the endotracheal tube in the trachea, the pilot balloon expands soon after you have filled the system cuff-pilot balloon with a few milliliters of air. In fact, as the cuff pushes on the trachea, also the trachea pushes against the cuff and, according to Pascal's principle (see Chap. 9), against the wall of the pilot balloon. You can see with your own eyes the much greater amount of air which is needed indeed to inflate the pilot balloon when the tube is in your hands or placed on a table; in this case, the push due to the *interaction* between the cuff and the tracheal wall is lacking.

14.2 Forces Against Motion: Normal Force and Friction

Now we have to deal with two types of forces that, in our everyday experience, oppose to the motion of objects. Both of them are involved, together with Newton's second law and gravity, in determining the balance of an object on an inclined plane and, in our "anesthesiologic" example, that of the patient's body in the Trendelenburg position.

14.2.1 Normal Force: Physics of the English Course

I don't know if the same occurs in other countries, but in Italy one of the first exercises of an entry level English course is, typically, "the pen is on the table." Physically speaking, if the pen is (motionless) on the table, it means that some kind of force must counterbalance the gravitational force which pulls down every object towards the center of the Earth (otherwise, the exercise would be "the pen is somewhere in the center of the Earth").

Actually, when a body lies on a surface, such as the pen on the table or a dresser on the floor, the surface exerts on the body the so-called *normal force* F_N, which completely *balances* (in the case of a horizontal surface such as the table or the floor) the gravitational force. This normal force is due to the deformation of the support surface and is always *perpendicular* to it (whence the term "normal," i.e., perpendicular).

If you are wondering whether, in the case of an object lying on a table, gravitational force and normal force are another example of action-reaction force pair, the answer is *no* (although this is a rather widespread belief). First, the action and reaction forces involved in Newton's third law never balance each other since they are applied to two different bodies interacting with one another (remember that *both* the rifle and the bullet move). Conversely, gravitational force and the normal force exerted by the table are both applied to the pen, which consequently remains in balance (i.e., motionless, which according to Newton's first and second laws means that no *net* force acts on it). Moreover, while action and reaction forces are parallel to each other, normal force is always perpendicular to the supporting surface; in the case of the pen on the table, gravitational force and normal force happen to be parallel since the table is horizontal, but it would not be so on an inclined plane, as we will see below.

14.2.2 Why Your Car Needs an Engine: Friction

According to Newton's first law (or to second law, if you prefer), if an object is moving at a constant speed, we can be sure that no net force acts on it. Why do we need to press on the gas pedal (and, accordingly, to consume fuel so that a driving force is applied to our car) in order to maintain, say, 120 km/h (about 75 mph) on the highway, though? Are Newton's laws valid only in physics textbooks?

Really, Newtonian mechanics does not apply to all phenomena of the real world. As mentioned, it does not apply to objects moving at very high speed (i.e., not negligible as compared with that of light), whose behavior is described by *special relativity* (see Chap. 21). Moreover, Newton's laws lose their validity for very small objects such as atoms and subatomic particles; in this case, we must rely on *quantum physics*. However, 120 km/h (or 75 mph) are absolutely negligible compared with the speed of light, and a car is definitely much bigger than an atom; accordingly, a car traveling on the highway is certainly comprised among those phenomena which are adequately described by Newton's laws.

The reason why our car needs a force to maintain a constant speed is that there is a force, which is called *friction*, that opposes to motion. Simply, when the car travels at a constant speed, the force generated by the engine exactly equals friction, so that no *net* force acts on the car (in full agreement with Newton's laws).

Generally speaking, whenever a force $\vec{F}$ tends to put in motion an object along a surface *S*, the surface itself exerts on the object a *frictional force* $\vec{F}_{\mathrm{f}}$ which is *parallel* to the surface and has an opposite direction to the (possible) motion, i.e., to the component of $\vec{F}$ which (possibly) causes the motion. The terms "possible" and "possibly" refer to the fact that, if the object is initially motionless, the force $\vec{F}$ will make it move only if the magnitude of the component of $\vec{F}$ parallel to the surface is *greater* than that of the frictional force (in fact, according to Newton's second law, you need a *net* force acting on the body in order to *accelerate* it). Consider, as an example, a heavy book lying on a table. If you push gently with a finger against it (parallel to the table), the book doesn't move due to the frictional force from the table. This kind of friction is known as *static friction*, that is, the frictional force between two bodies which are not moving relative to each other. If you push harder (so that the force you apply is *greater* than the static frictional force), the book starts to slide along the table, and you need to keep pushing in order to *equal* the so-called *kinetic friction* (i.e., the frictional force acting between moving objects) and maintain the book in uniform rectilinear motion, as in the case of the car on the highway. According, once again, to Newton's second law, if the force you apply is greater than the kinetic frictional force, the book accelerates.

The maximum magnitude of *static* friction for an object lying on a surface *S* is:

$$F_{\mathrm{f}} = \mu F_{\mathrm{N}} \tag{14.3}$$

where F_{N} is the magnitude of the normal force exerted by the surface *S* on the object (see Sect. 14.2.1) and μ is the so-called *coefficient of static friction*, which depends on the characteristics of both the object and the surface; for example, μ is 1.0 for rubber on concrete (i.e., the maximum static frictional force is equal to the normal force exerted by the supporting surface), 0.2–0.6 for wood on metals, and is very low (about 0.03) for steel on ice, as ice skaters know very well!

14.3 Gravity Vs. Friction: Safety in the Trendelenburg Position

A steep (up to 35°-45°) *Trendelenburg position*, i.e., head down tilt, is often required in procedures such as laparoscopic (especially robotic) urologic and gynecologic surgery [6–8]. Apart from the well-known respiratory, hemodynamic, and cerebrovascular effects and the other possible complications (including soft tissue edema and ocular injury) [7, 9–11], a major concern of this position is the slipping of the

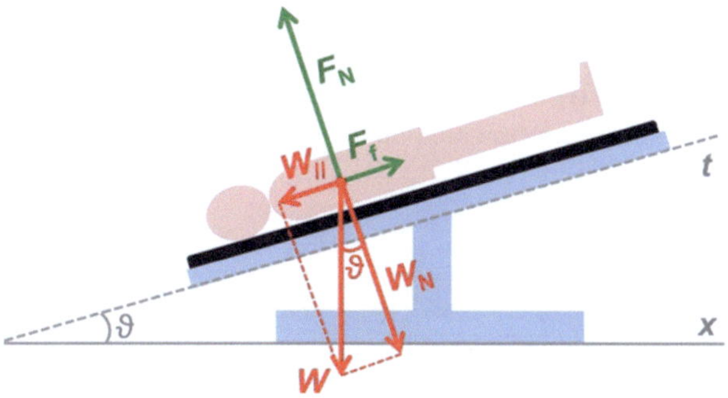

Fig. 14.2 The forces involved in the balance of a patient in the Trendelenburg position. *W* patient's weight, resolved in its components perpendicular (W_N) and parallel ($W_{//}$) to the operating table; F_N normal force; F_f frictional force

patient along the operating table (under the action of gravity); this may even cause the patient to fall off the table or, if some restraining method such as shoulder blocks, wristlets, braces, or flexed knees is used, may lead to nerve injury (e.g., to the brachial plexus) due to compression or stretch [6, 7, 9].

Evidently, the patient does not slip and, accordingly, does not risk falling or having peripheral nerve injury (if restrained) as long as the frictional force exerted by the table is sufficient to balance the effects of gravity. This depends, intuitively, on the "degree" of tilt, expressed as the width of the angle ϑ between the table and a line parallel to the floor (Fig. 14.1). Which is, therefore, the maximum allowable tilt in order to prevent sliding?

An old article stated that some restraining method is needed to keep the patient in a 40° Trendelenburg position [12]. Let's give a closer look to the forces involved (and to their relationship with the tilt angle) in order to provide a "physical" explanation for this statement and to make some other interesting considerations.

As shown in Fig. 14.2, three forces (all of them applied *to* the patient) must balance each other to keep the patient in the Trendelenburg position without using blocks or other fasteners: the patient's weight *W* (i.e., the gravitational force), the normal force F_N exerted by the operating table as a supporting surface (see Sect. 14.2.1), and the frictional force F_f due to the tendency of the patient to slip down the table under the effect of gravitational force. The patient's weight can be resolved into two components (see Chap. 13); the component W_N, perpendicular (or "normal") to the table surface (*t*), is counterbalanced by the normal force F_N, while the component $W_{//}$, parallel to *t*, tends to slide the patient down the table.

Remembering the simple notions of trigonometry discussed in Chaps. 1 and 13 (see Fig. 1.2 or Box 13.1) for quick reference) and noting that the angle ϑ between the table and the floor (*x*) is equal to the angle between the vectors *W* and W_N (since *W* is perpendicular to *x* and W_N is perpendicular to *t*), the magnitude of the parallel component is:

$$W_{//} = W \sin \vartheta \tag{14.4}$$

or, after substituting for *W* from Eq. (14.2):

$$W_{//} = m\,g \sin \vartheta \tag{14.5}$$

where m is the mass of the patient (what we usually call "weight") and g is the free-fall acceleration.

Similarly, the magnitude of the maximum static frictional force which opposes to sliding can be calculated as follows (keep in your mind Eq. 14.3 and, again, Eq. 14.2 and what discussed in Sect. 1.3.1 or in Box 13.1):

$$F_{\mathrm{f}} = \mu F_{\mathrm{N}} = \mu W \cos \vartheta = \mu m g \cos \vartheta \tag{14.6}$$

where μ is the coefficient of static friction for the patient's skin to the table surface and F_{N} is the normal force.

Note that, for a ϑ value between 0° and 90°, sin ϑ increases with the angle width, while cos ϑ decreases. In particular, sin ϑ is lower than cos ϑ for $\vartheta < 45°$, sin ϑ is higher than cos ϑ for $\vartheta > 45°$, and sin ϑ = cos ϑ for $\vartheta = 45°$ (see Chap. 1). Accordingly, $W_{//}$ increases, and F_{f} decreases, with the width of ϑ.

As mentioned, the patient remains in balance (i.e., no net forces act on him/her and, accordingly, he/she doesn't *accelerate* downward or, if restrained, doesn't push or pull against restraints) as long as it is:

$$F_{\mathrm{f}} = W_{//} \tag{14.7}$$

According to Eqs. (14.6) and (14.5) it must be, therefore:

$$\mu m g \cos \vartheta = m g \sin \vartheta \tag{14.8}$$

or, after dividing both sides by mg:

$$\mu \cos \vartheta = \sin \vartheta \tag{14.9}$$

This applies, of course, to any object lying on an inclined plane (or *ramp*). Consider, as an example, a wooden crate (perhaps full of bottles of champagne) on a downhill concrete road. Assuming a coefficient of static friction μ for wood on concrete of about 0.6, the crate will not slide down (if no one pushes against it!) on a road with a slope lower than 30°; in fact, cos 30° = 0.866 and sin 30° = 0.5 ≅ 0.6 × 0.866, that is Eq. (14.9) for $\mu = 0.6$ and $\vartheta = 30°$. The precious bottles will be certainly safe on a 10° descent, unless the road is, say, snowy; in this case, a slope of 8° would be enough to make the crate slip (μ is just 0.14 for wood on wet snow, cos 8° = 0.99, sin 8° = 0.139 ≅ 0.14 × 0.99).

From Eq. (14.9) we can also say that the coefficient of static friction μ between two substances is the *trigonometric tangent* (see Chap. 1, Sect. 1.3.1) of the angle of inclination ϑ of a ramp made of one of the two substances for which an object made of the other substance, positioned on the ramp, *starts* to slide (because the forces acting on it are balanced and, according to the Newton's first and second law discussed above, it is "free" to move with constant speed). In fact, Eq. (14.9) can be written as:

$$\mu = \frac{\sin \vartheta}{\cos \vartheta} = \tan \vartheta \tag{14.10}$$

It should be noted that the coefficient of friction between two substances is not exactly constant (furthermore, Eq. 14.3 itself is only an approximation, as clearly explained by the famous Nobel laureate Richard Feynman in its landmark textbook "*The Feynman Lectures on Physics*" [13]).

In the case of a patient on the operating table, the issue is much more complex. The claim that some restraining method is needed to keep the patient at a 40° tilt [12] might make sense provided you do not rely too much on this value as a safety limit. Assuming as coefficient of static friction the median value ($\mu = 0.73$) found between in vivo human skin and an aluminum sample in a study a few years ago [14], the patient would slip (along an aluminum table) with $\vartheta \geq 36°$ ($\cos 36° = 0.81$; $\sin 36° = 0.59 \cong 0.73 \times 0.81$). However, the top surface of operating tables is usually not made of aluminum. Moreover, different operating tables may exert different frictional forces, and the placement of a drape (e.g., a disposable bedsheet) between the table and the patient may result in a significant reduction in the coefficient of friction [15]. Most remarkably, a lot of factors may affect the frictional properties of human skin, including age, anatomical site, presence of hair, hydration, sweating, temperature, and ambient humidity, to name but a few [14, 16, 17]. Thus, it is very difficult (if not impossible) to exactly predict the maximum tilt angle allowable to prevent sliding; it might be much lower than 35°–40°, depending on several factors.

A possible way to increase safety in the Trendelenburg position, allowing a higher tilt angle without resorting to restraining methods (or at least preventing nerve injury due to compression or stretch caused by them) is the use of anti-slip devices, such as pads or mattresses, which increase the coefficient of friction [6, 18, 19]. However, further studies are needed to better understand their usefulness and, in particular, to what extent they safely increase the maximum tilt allowed.

Finally, one might think that obese patients (simply because they "weigh" more, i.e., have a higher body *mass*) are at higher risk of downward slipping in the Trendelenburg position, as some authors stated [19, 20]. Conversely, it is clear from Eqs. (14.8) and (14.9) that mass (which in fact disappears from Eq. 14.8 since it is present in both its sides) equally contributes to both the force pushing down (i.e., the parallel component of the patient's weight) and the frictional force which opposes to it; accordingly, patient's body mass does not affect (by itself) the risk of slipping, unless we consider, for some reason, the coefficient of static friction as a function of body mass. Indeed, since the patient is not a wooden crate but a (complex) biological system, it cannot be excluded that obesity may affect the frictional properties of skin or of the whole patient's body (remember that the skin's coefficient of friction may vary among different anatomical sites). However, this may not necessarily reduce the coefficient of static friction, but even *increase* it, leading to a reduced risk of sliding. Consistently, although Klauschie et al. [19] found that body mass did not significantly affect the extent of patient's shift along an anti-slip device (pink egg-crate foam) in the Trendelenburg position, a trend towards a reduced shift was noted in obese patients. It is also true, however, that while it is not the increased mass per se which may increase the risk of slipping (and, accordingly, of nerve injury or falling) in the Trendelenburg position, especially once a "final" position has been established, obese patients may undergo more "mass shifts" than normal-weight patients *during* the increase in tilting, and these shifts can contribute to start movement along the operating table plane [20].

Future well-designed studies about these topics are desirable and would be very interesting, but are rather difficult to perform rigorously. Meanwhile, it is advisable not to exceed with tilt, especially since recent findings suggest that most robotic procedures can be performed safely (from a surgical standpoint) with a Trendelenburg angle of less than 30° [21, 22].

References

1. Styer DF. Relativity for the questioning mind. Baltimore, MD: The Johns Hopkins University Press; 2011.
2. Durell CV. Readable relativity. New York: Dover Publications; 2003.
3. Einstein A. Relativity: the special & the general theory, 100th anniversary edition. Princeton, NJ: Princeton University Press; 2015.
4. Gribbin J. Einstein's masterwork: 1915 and the general theory of relativity. Icon Books; 2015.
5. Fischer K. Relativity for everyone. How space-time bends. Cham: Springer; 2015.
6. Wechter ME, Kho RM, Chen AH, Magrina JF, Pettit PD. Preventing slide in Trendelenburg position: randomized trial comparing foam and gel pads. J Robot Surg. 2013;7(3):267–71.
7. Hsu RL, Kaye AD, Urman RD. Anesthetic challenges in robotic-assisted urologic surgery. Rev Urol. 2013;15(4):178–84.
8. Shveiky D, Aseff JN, Iglesia CB. Brachial plexus injury after laparoscopic and robotic surgery. J Minim Invasive Gynecol. 2010;17(4):414–20.
9. Martin JT. The Trendelenburg position: a review of current slants about head down tilt. AANA J. 1995;63(1):29–36.
10. Gainsburg DM. Anesthetic concerns for robotic-assisted laparoscopic radical prostatectomy. Minerva Anestesiol. 2012;78(5):596–604.
11. Gkegkes ID, Karydis A, Tyritzis SI, Iavazzo C. Ocular complications in robotic surgery. Int J Med Robot. 2015;11(3):269–74.
12. Hewer CL. Maintenance of the Trendelenburg position by skin friction. Lancet. 1953;1(6759):522–4.
13. Feynman RP, Leighton RB, Sands M. The Feynman lectures on physics: the new millennium edition. New York: Basic Books; 2011.
14. Veijgen NK, Masen MA, van der Heide E. Variables influencing the frictional behaviour of in vivo human skin. J Mech Behav Biomed Mater. 2013;28:448–61.
15. Nakayama JM, Gerling GJ, Horst KE, Fitz VW, Cantrell LA, Modesitt SC. A simulation study of the factors influencing the risk of intraoperative slipping. Clin Ovarian Other Gynecol Cancer. 2014;7:24–8.
16. Veijgen NK, van der Heide E, Masen MA. A multivariable model for predicting the frictional behaviour and hydration of the human skin. Skin Res Technol. 2013;19(3):330–8.
17. Derler S, Gerhardt LC. Tribology of skin: review and analysis of experimental results for the friction coefficient of human skin. Tribol Lett. 2012;45:1–27.
18. Greenberg JA. The pink pad-Pigazzi patient positioning system™. Rev Obstet Gynecol. 2013;6(2):97–8.
19. Klauschie J, Wechter ME, Jacob K, Zanagnolo V, Montero R, Magrina J, Kho R. Use of anti-skid material and patient-positioning to prevent patient shifting during robotic-assisted gynecologic procedures. J Minim Invasive Gynecol. 2010;17(4):504–7.
20. Dybec RB. Intraoperative positioning and care of the obese patient. Plast Surg Nurs. 2004;24(3):118–22.

21. Gould C, Cull T, Wu YX, Osmundsen B. Blinded measure of Trendelenburg angle in pelvic robotic surgery. J Minim Invasive Gynecol. 2012;19(4):465–8.
22. Aggarwal D, Bora GS, Mavuduru RS, Jangra K, Sharma AP, Gupta S, Devana SK, Parmar K, Kumar S, Mete UK, Singh SK. Robot-assisted pelvic urologic surgeries: is it feasible to perform under reduced tilt? J Robot Surg. 2020; https://doi.org/10.1007/s11701-020-01139-7. Online ahead of print

Part VI

Thermology and Inhalational Anesthesia: The Physics of Vaporizers

Physics in a Vaporizer: Saturated Vapor Pressure, Heat of Vaporization, and Thermal Expansion

15

Contents

Where's Physics

Daily life	Water boiling in your kitchen Pressure cooker Absolute and relative humidity A drink left in a glass in the garden (on a windy day) Clothes drying in the wind Pots (and handles) Thermal excursion in the desert Train tracks, bridges, and thermostats

A. Pisano, *Physics for Anesthesiologists and Intensivists*,
https://doi.org/10.1007/978-3-030-72047-6_15

Physics involved	State changes of matter (particularly vaporization) Saturated vapor pressure Heat of transformation and evaporative cooling Heat transfer Thermal conductivity Specific heat Dalton's law (see Chap. 4) Thermal expansion
Clinical practice	Anesthetic vaporizers (how they work, why are so heavy, what makes them quite reliable, how they adapt to the ambient conditions) Humidification of inspired gases in the airways Cylinders containing liquefied gases

15.1 Why a Vaporizer Is Not Exactly a "Vaporizer": Saturated Vapor Pressure and Volatility

A thorough coverage of the different types of vaporizers, the way they work, and their advantages and drawbacks is beyond the scopes of this book and can be found elsewhere, including most anesthesia textbooks [1–4]. This chapter is limited to pointing out and discussing the many physical principles (some of which have already been mentioned in previous chapters) which are the basis of the operation and design of vaporizers, with special reference to the so-called *variable bypass vaporizers*, i.e., those most commonly used for the administration of halogenated anesthetics such as isoflurane and sevoflurane (but not desflurane). However, since we are going to talk about physics and halogenated anesthetics (although exclusively with regard to the devices through which they are administered) in the same chapter, you may find it interesting to take a look at Box 15.1 (for a brief digression) before starting with the first basic concept we need to discuss in order to understand the physics of vaporizers, namely saturated vapor pressure.

Box 15.1: Just Out of Curiosity: Halogenated Anesthetics and Quantum Physics

You may have happened to read somewhere that the mechanism of action of halogenated anesthetics, which is not yet fully understood, could have something to do with *quantum physics* [5–7]. This is closely linked to the existence of a set of theories referred to as "quantum consciousness," according to which our consciousness would have a quantomechanic basis. Quantum physics, or *quantum mechanics*, one of the three pillars of modern physics (the other two being thermodynamics, which is discussed in Chap. 10, and Relativity, which is briefly mentioned several times in the book) [8], is the physics of very small objects such as atoms and subatomic particles, and is full of "oddities." For example, "particles" sometimes behave as particles, while other times as waves (and "waves" do the same). The more the accuracy with which we know the position of a particle, the less the certainty about its speed (the same applies to other couples of "complementary" physical quantities, according to the so-called *Heisenberg's uncertainty principle*). Any quantum system can exist in a *superposition* of quantum states, at least until you *observe* it; in other words, if I ask my wife where are my green socks, and she answers "maybe in the wardrobe or in your bedside table," if we were talking about a quantum system they would be *both* in the wardrobe *and* in the bedside table until I find them! Maybe, the most incredible aspect of quantum physics is, however, its "non-locality" (something Einstein couldn't digest at all); for example, two particles which have to be considered as a unique quantum system remain correlated (or, as they say, *entangled*) to each other even if they will subsequently be separated by millions of kilometers; if the quantum state of one of them changes, also that of the other one will change *instantaneously*. The best known of the abovementioned quantum consciousness theories is the so-called *orchestrated objective reduction* (Orch-OR) theory developed by Roger Penrose, a famous Nobel laureate in physics, and the anesthesiologist Stuart R. Hameroff. According to this theory, our consciousness would be the result of the quantum *superposition* of two different states of tubulin (the protein of which the microtubules contained in neurons are made) and to the quantum *entanglement* among tubulin molecules of different neurons in our brain (in practice, our neurons would be connected to each other "wirelessly") [9]; halogenated anesthetics would be able to interact with this "quantum system." If you are quite (or very) perplexed by all this, don't worry; it is absolutely normal! And this is one of the two reasons why I didn't devote a chapter to this topic (dealing with this subject with the same depth of the others in this book would have been too long and complex). The second (and most important) reason is that this theory, although intriguing, lacks experimental confirmation and does not meet large consensus among quantum physicists and neurobiologists worldwide [10].

15.1.1 Saturated Vapor Pressure and Boiling Point

Any liquid, in an enclosed container, evaporates (i.e., vaporizes *from its surface*) until it reaches thermodynamic equilibrium, that is until the number of molecules which have a kinetic energy high enough to escape from the liquid phase (to form vapor) equals that of molecules returning to the liquid phase. The pressure of the vapor in these circumstances is called *saturated vapor pressure* (SVP) and represents, evidently, the maximum vapor pressure possible under these conditions. Since, as discussed in Chap. 2 (see Box 2.2), the kinetic energy of molecules increases with temperature (or better, a higher temperature precisely *means* a higher average kinetic energy of molecules), it goes without saying that the saturated vapor

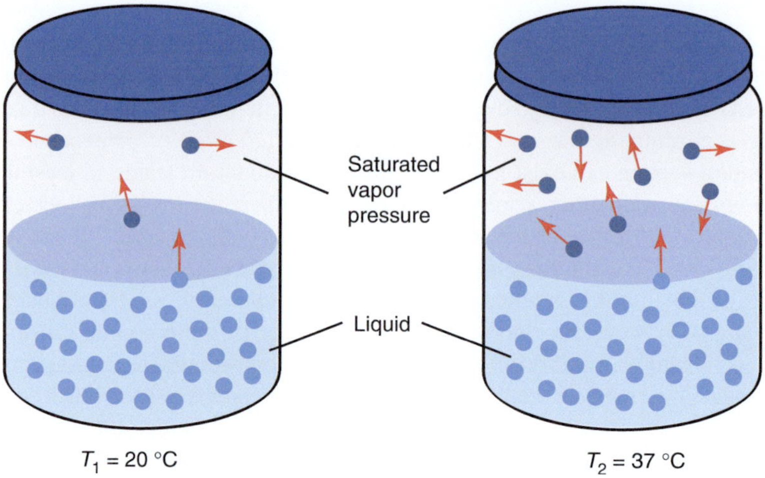

Fig. 15.1 Saturated vapor pressure (SVP) can be defined as the pressure exerted by a vapor which is in thermodynamic equilibrium with its condensed phase in an enclosed space. SVP is a function of temperature; in fact, a larger number of molecules has sufficient kinetic energy to leave the liquid phase at any temperature $T_2 > T_1$. Red arrows represent the speed vectors of each vapor molecule

pressure of any substance *increases with temperature* (see Fig. 15.1). For example, the SVP of water is 17.5 mmHg (or 2.3 kPa) at 20 °C (or 68 °F). When you put some water, whose initial temperature is 20 °C (68 °F), to boil in a pot, the *partial pressure* of water vapor just above the liquid surface, which is initially about 17.5 mmHg (i.e., equal to the SVP of water at that temperature), increases progressively as temperature rises. As mentioned in Chap. 2 (see Box 2.1), water will boil when its vapor pressure will equal the external atmospheric pressure. In these conditions, in fact, water vapor has a sufficient pressure to form bubbles within the bulk of liquid, and vaporization turns from a "surface phenomenon" to a "volume phenomenon." The temperature at which this occurs is called *boiling point*. Evidently, the boiling point of a given substance varies according to the surrounding environment pressure. For example, the SVP of water is equal to 760 mmHg (or 101.3 kPa), that is the atmospheric pressure at sea level, at a temperature of 100 °C (212 °F); this is, accordingly, the temperature at which water boils at sea level (or its *normal boiling point*). In a pressure cooker, the "surrounding environment pressure" becomes higher and higher as water vapor accumulates in an enclosed space; thus, the boiling point of water increases (up to 120 °C or 248 °F), and foods are cooked faster. As discussed in Chap. 2 (see Box 2.1), the opposite occurs in the high mountains.

The notions discussed above provide an explanation for what is stated in Chap. 4 about air humidity and the humidification of inhaled air in the airways (see Box 15.2). Also the issue, discussed in Chap. 2 (Sect. 2.4), of a cylinder containing a liquefied gas, whose content cannot be "measured" through a pressure gauge (which will always measure nothing but the SVP, whatever the remaining volume of liquid), is now certainly clearer.

Box 15.2: Saturated Vapor Pressure and (Air) Humidity
Atmospheric humidity can be defined roughly as the "amount" of water vapor in the air (*absolute humidity*). As in the example of the pot of water (Sect. 15.1.1), the higher the temperature, the higher the partial pressure of water vapor in the air (consider the pot of water as the sea, lakes, and so on in a certain region, and your kitchen as the surrounding atmosphere). Accordingly, *absolute humidity increases with temperature*. However, humidity is usually expressed as the ratio between the actual amount of water vapor and the maximum amount of water vapor which air can "hold" at a given temperature (*relative humidity*). This "maximum amount" is, evidently, the saturated vapor pressure (SVP) of water at that temperature (similarly, the maximum humidification of inhaled air in the airways corresponds to the SVP of water at 37 °C, or 98.6 °F, as mentioned in Chap. 4). Since the SVP increases with temperature, relative humidity *decreases* with temperature (for a fixed "absolute amount" of water vapor in the air, i.e., if absolute humidity does not change); that's why when you turn on the heating at your home the air gets "dry" (whereas humidifiers, providing a source of water, increase absolute humidity and, thus, prevent an excessive reduction in relative humidity).

15.1.2 Volatility of Halogenated Anesthetics and the "Trick" of the Variable-Bypass Vaporizer

The saturated vapor pressure of a liquid at a given temperature (usually 20 °C, or 68 °F) is a measure of the *volatility* of such liquid, i.e., of its tendency to vaporize; the higher the SVP at 20 °C (68 °F) of the liquid, the more quickly it vaporizes (at that temperature).

The term "vaporizer" seems to suggest that these devices *cause* the vaporization of halogenated anesthetics. Conversely, these drugs vaporize, *spontaneously*, very easily (for no other reason they are also called "volatile anesthetic agents"). As shown in Table 15.1, in fact, all halogenated anesthetics have an SVP at 20 °C much

Table 15.1 Approximated values of saturated vapor pressure (SVP) at 20 °C of halogenated anesthetics. SVP of water is also reported for comparison

	SVP at 20 °C (68 °F)
Water	17.5 mmHg (2.3 kPa)
Sevoflurane	157.5 mmHg (21 kPa)
Enflurane	172 mmHg (23 kPa)
Isoflurane	240 mmHg (32 kPa)
Halotane	244 mmHg (32.5 kPa)
Desflurane	669 mmHg (89 kPa)

higher than water [11]. Sevoflurane, for example, which is about nine times more volatile than water (and nevertheless is the less volatile among the halogenated anesthetics), has an SVP at 20 °C of 157.5 mmHg (or 21 kPa). This would be, therefore, the pressure of sevoflurane vapor in an enclosed space containing the anesthetic in liquid form, in equilibrium with its vapor, at the temperature of 20 °C. If a fresh gas mixture flowed through such a *vaporizing chamber*, coming into contact with sevoflurane vapor, the concentration of sevoflurane in the outlet mixture would be, according to Dalton's law (see Chap. 4) and remembering that the atmospheric pressure at sea level is 760 mmHg:

$$\frac{157.5\,\text{mmHg}}{760\,\text{mmHg}} \times 100 = 20.7\% \qquad (15.1)$$

It is evident that sevoflurane does not need at all a "vaporizer" to vaporize; in fact, it vaporizes even too much by itself! The issue is how to reduce this very high concentration to the clinically adequate (and safe) values of around 1–2% [12], and how to set the desired concentration with a simple action (that is, by turning the concentration dial).

In a common *variable bypass vaporizer* (see Fig. 15.2), the concentration of anesthetic delivered to the patient depends on how much fresh gas enters the vaporizing chamber (i.e., the space containing the liquid agent in equilibrium with its vapor) as compared with how much is diverted into a "bypass"; the concentration dial simply sets this *splitting ratio*. Thus, the maximum concentration possible (i.e., about 21%) is more or less diluted by the fresh gas that flows into the bypass, in order to obtain the desired concentration. These vaporizers are, obviously, agent-specific, since the splitting ratio needed to get a certain concentration depends on the SVP. For example, isoflurane is much more volatile than sevoflurane (see Table 15.1); accordingly, as compared to sevoflurane, a smaller fraction of the fresh gas flow needs to enter the vaporizing chamber to obtain the same concentration.

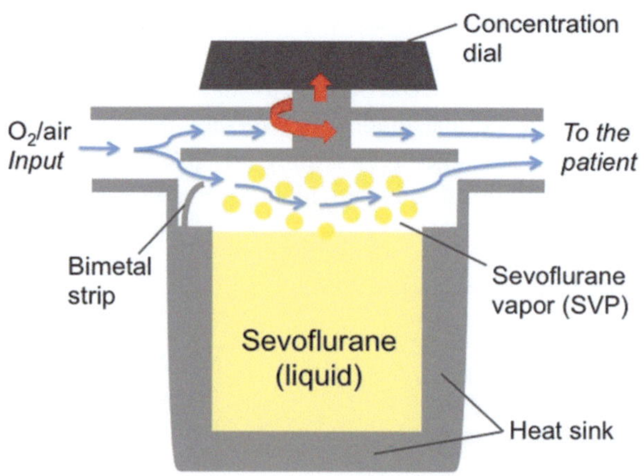

Fig. 15.2 Schematic representation of a variable bypass vaporizer for sevoflurane (see text). The *heat sink* (see Sect. 15.2.3) and the *bimetal strip* (see Sect. 15.3) provide temperature stabilization and compensation, respectively. *SVP* Saturated vapor pressure

15.1.3 Why Desflurane Needs a Special Kind of Vaporizer

As shown in Table 15.1, desflurane is the most volatile among halogenated anesthetics. Clearly, the higher the SVP of a liquid, the lower its boiling point. Since the SVP at 20 °C of desflurane (669 mmHg or 89 kPa) is already very close to the atmospheric pressure at sea level (760 mmHg or 101.3 kPa), its normal boiling point is only slightly higher than 20 °C (i.e., around 23 °C or 73 °F). Accordingly, desflurane could boil at room temperature (including that of an operating room), especially at higher altitudes (see Chap. 2), leading to hazardous doses delivered to the patient. Moreover, a very high bypass flow would be needed in order to dilute the concentration of desflurane at the desired values since the concentration leaving the vaporizing chamber at 20 °C would be 100 x (669/760) = 88%! Mainly for these reasons, the common variable bypass vaporizers are unsuitable for the administration of desflurane, that needs a special kind of vaporizer in which the anesthetic is heated at 39 °C (about 102 °F), reaching a vapor pressure of about 1500 mmHg (or 2 atm), and its inlet into the fresh gas flow is controlled through an electronically operated system consisting of a pressure regulating valve and a pressure transducer [1, 3, 4, 13–15].

15.2 Why Vaporizers Are So Heavy: Heat of Transformation and the Need for Temperature Stabilization

If you have ever replaced the (variable bypass) vaporizer in your anesthesia workstation, you well know how much these devices weigh (more or less like a 2-year-old child!). As discussed below, also this has to do with physics, in particular with the *state* (or *phase) change* which occurs in the vaporizer and the consequent need to keep its temperature constant in order to maintain its accuracy.

15.2.1 Some Notes About the State Changes of Matter

In Sect. 15.1.1, we reminded why the boiling point of a liquid depends on the surrounding pressure (as already mentioned in Chap. 2). This applies, actually, to every *state change* of matter; the temperature at which any substance melts, vaporizes, etc. is characteristic of that substance, but varies with pressure. Accordingly, it makes no sense to say that the *melting point* of lead is 328 °C (or 622 °F) without specifying "at normal (i.e., sea-level) atmospheric pressure" (although in the current language, also among physicists, the term "melting point" conventionally refers to this pressure). Note that the temperature at which a state change occurs (under a given pressure) is always the same of the reverse transformation; for example, the boiling point of water at a given pressure is equal to its "condensing point" at that pressure, as well as its melting point is equal to its "freezing point." In fact, these are the temperatures at which two states (e.g., liquid and vapor, solid and liquid)

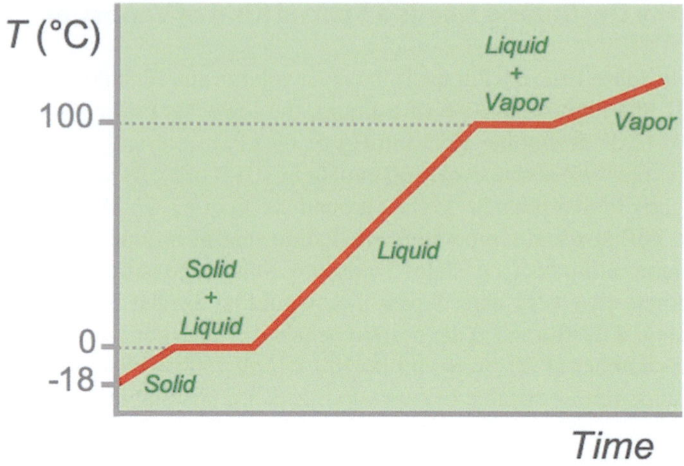

Fig. 15.3 Imagine to warm an ice cube, whose initial temperature T is −18 °C, in an enclosed container. When its temperature reaches the melting (or freezing) point of water, that is 0 °C, ice begins to melt. Temperature remains constant (i.e., 0 °C), despite you continue to provide heat, until *all* water has turned from solid to liquid. Only then does the temperature rise again. At the boiling point of water (100 °C at the sea level), liquid starts to boil, and temperature remains again constant (i.e., 100 °C) until *all* liquid has changed into vapor. Only then does the temperature start to raise again (assuming you continue to provide heat). This is because state changes of matter need energy, under the form of heat of transformation (see text)

coexist. Accordingly, they remain constant until *all* the amount of substance has been converted into another state (see Fig. 15.3). Consider again some water that is boiling in a pot (at sea level). We had put the water on the stove in order to transfer some *heat* to it and, hence, increase its temperature (see Chap. 10) up to its boiling point, i.e., 100 °C; as long as some liquid water remains in the pot, we can be sure that its temperature still remains 100 °C *although we continue to provide heat.* Contrary to the heat supplied initially, which causes an increase in temperature, this heat is needed to free the water molecules from their clusters, i.e., to overcome the *cohesive forces* between them (see Chap. 6). In general, every state change of matter is accompanied by an *energy transfer* (as absorbed or released heat); the amount of energy which needs to be transferred so that a unit mass of a given substance completely undergoes a state change is called *heat of transformation* (or *latent heat*). In particular, the *heat of vaporization* (also referred to as *latent heat of vaporization*) is the amount of energy which must be supplied to a unit mass of liquid in order to completely convert it into vapor.

15.2.2 Evaporative Cooling and Accuracy of Vaporizers

There is no doubt that, during inhalational anesthesia, the liquid anesthetic agent (say, sevoflurane) contained in the vaporizer is progressively converted into vapor.

In fact, as the fresh gas mixture flows through the vaporizing chamber, it carries away some amount of vapor. Consequently, other sevoflurane molecules leave the liquid phase in order to restore the equilibrium (i.e., to maintain the vapor pressure at the value of the SVP). According to what discussed above, this process requires energy (heat of vaporization). Since, unlike the example of the pot of water on the stove, no one supplies heat from the outside, the anesthetic itself must provide this heat and, hence, it tends to cool (*evaporative cooling*) [1]. Accordingly, its SVP, which depends on temperature (as discussed in Sect. 15.1.1), decreases. Evaporative cooling of halogenated anesthetics is most pronounced when high fresh gas flows are used. In fact, the higher the flow of fresh gas through the vaporizing chamber, the faster the vapor has to be replaced from the liquid reservoir and, therefore, the more rapidly the temperature (and, as a consequence, the SVP) will drop. For the same reason, a liquid in an open container exposed to the wind (such as a soft drink left in a glass in the garden) is always at a temperature lower than the ambient temperature. Similarly, clothes will dry quickly (i.e., water will evaporate quickly) on a windy day, but they will appear cold.

Since the splitting ratio which must correspond to every mark on the concentration dial in order to obtain the desired anesthetic concentration is calibrated on a certain value of SVP of the anesthetic agent (e.g., that at 20 °C), the accuracy of variable bypass vaporizers decreases progressively as the temperature (and, accordingly, the SVP) of the anesthetic agent falls, leading to actual concentrations lower than those you set. For this reason, temperature must be kept as constant as possible in variable bypass vaporizers (we will see how in the next section).

15.2.3 Temperature Stabilization (Heat Sink): Specific Heat and Thermal Conductivity

In modern variable bypass vaporizers, *temperature stabilization* is achieved thanks to the presence of a *heat sink*, namely a large mass of a metal alloy with high *specific heat* and high *thermal conductivity* [4]. As discussed in Chap. 10, the specific heat c of any substance is the amount of heat Q which must be provided to the unit mass of that substance in order to increase its temperature by 1 °C. In equation form:

$$c = \frac{Q}{m\Delta T} \tag{15.2}$$

where m is mass in grams (g) and ΔT is the temperature change (note that Eq. (15.2) is simply Eq. (10.7), already encountered in Chap. 10, written in a different form). Since heat is a form of energy, whose unit in the International System of Units (SI) is the joule (J), the SI unit for specific heat is J/g K (which, in this case, is the same as J/g °C since a temperature *change* has the same value in both kelvin [K] and Celsius degrees [°C]). As we saw in Chap. 10 (Sect. 10.2), a pot should be made of materials with low specific heat (e.g., steel, which has a specific heat of about 0.47 J/g °C), so that not too much heat is needed to increase its temperature enough to cook your meals, while its handles should be preferably made of materials with higher specific heat (e.g., plastic, which has a specific heat more than three times greater than steel) as to not get burned when you touch them (in practice, the same heat provided to 1 g of steel or to 1 g of plastic causes a temperature increase of steel which is more than threefold that of plastic).

If we refer to an object having any mass *m*, we define the *heat capacity* (or *thermal capacity*) *C* of this object as the heat which must be provided to it in order to increase its temperature by 1 °C (or, which is the same, the heat which it must *release* in order to reduce its temperature by 1 °C). This heat is evidently (putting $\Delta T = 1$ °C in Eq. 15.2):

$$Q = C = cm \tag{15.3}$$

Hence, heat capacity of an object is the product of its mass and the specific heat of the substance of which it is made. As you may have noticed, also Eq. (15.3) is nothing more than Eq. (10.6); here, we are simply doing the reverse reasoning as compared with Chap. 10. In that chapter, we had also already said that the absence of the sea is one of the reasons for the wide thermal excursion between night and day in the desert. The sea, in fact, has a large heat capacity *C*, due both to the high specific heat of water ($c = 3.9$ J/g K for seawater) and to the undoubted immensity of its mass; accordingly, as mentioned, it can store a lot of heat during the sunny hours (with small increases in its temperature), and "return" it overnight.

Heat can be transferred from one object to another by *conduction*, i.e., through contact with a solid slab; *convection*, i.e., through a fluid such as air or water; or *irradiation*, i.e., by means of electromagnetic waves (see Chap. 16). The amount of heat *Q* which is transferred *by conduction* over the time *t* (i.e., the so-called *heat conduction rate Q*/*t*) between the two faces of a slab with thickness *l* and face area *A*, whose temperatures are T_1 and T_2, respectively, can be calculated according to the following equation:

$$\frac{Q}{t} = kA\frac{T_1 - T_2}{l} \tag{15.4}$$

where *k* is a constant, called *thermal conductivity*, which depends on the material of which the slab is made; the higher the thermal conductivity, the faster the heat transfer. Of course, according to the second law of thermodynamics (see Chap. 10), heat will be always transferred from the face with higher temperature to that with lower temperature.

Due to its large heat capacity, the heat sink contained in the vaporizer is capable of releasing a lot of heat with a small decrease in its temperature. Moreover, due to its high thermal conductivity, it readily transfers heat to the anesthetic. Accordingly, the heat sink (to which the considerable weight of vaporizers is largely due) minimizes the fall in temperature (and, accordingly, in SVP) of the anesthetic due to evaporative cooling and, hence, preserves the accuracy of the vaporizer.

15.3 Thermal Expansion: Train Tracks, Thermostats and Temperature Compensation in Vaporizers

In addition to temperature stabilization, modern variable bypass vaporizers are also *temperature compensated*, i.e., the splitting ratio varies with temperature in order to keep the anesthetic concentration delivered constant [1–4]. In fact, ambient temperature in the operating room affects the SVP of the anesthetic agent and, accordingly, the share of fresh gas which needs to be directed into the vaporizing chamber in order to get a certain concentration. For example, if the operating room temperature is kept at around 17 °C (such as during interventions requiring deep hypothermic circulatory arrest), the SVP of sevoflurane is lower than that at 20 °C and, accordingly, a *higher* proportion of fresh gas needs to enter the vaporizing chamber as compared to that needed to get the same concentration at 20 °C.

Temperature compensation is also based on a physical phenomenon: *thermal expansion*. You surely know that, until a few years ago, between two consecutive train tracks there was always a little space that allowed them to lengthen, during the warmer seasons, without clashing each other; otherwise, they could be dangerously deformed (modern high-speed train tracks use a different design for this purpose). The same applies to the different sections of a bridge. In fact, when the temperature of a metal rod of length l is raised by an amount ΔT, its length will increase by an amount Δl, according to the equation:

$$\Delta l = l\,\alpha\,\Delta T \tag{15.5}$$

where α is the so-called *coefficient of linear expansion*, a constant depending on the material.

The common liquid-in-glass clinic thermometers are based on the fact that, as temperature rises, liquids such as mercury (now banned in many countries due to toxicity and environmental hazard) or the currently used *galinstan* (the registered brand name of a metal alloy of gallium, indium, and tin with a melting point around −19 °C, which is therefore liquid at room temperature) expand to a greater extent than their glass container.

Another way to take advantage of the different coefficients of linear expansion of two different materials is to weld together two metals, for example, copper and steel. Such a *bimetal strip* (Fig. 15.4) will bend as temperature varies, due to the

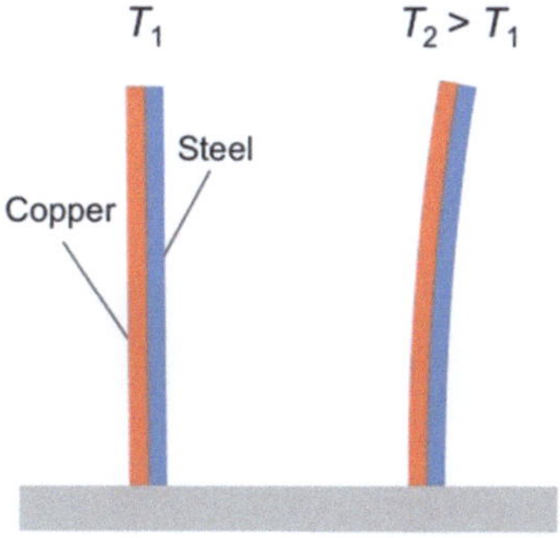

Fig. 15.4 A bimetal strip bends (evidently, towards the side of the metal with the lower coefficient of linear expansion α) as its temperature increases. α is 17×10^{-6}/°C for copper and 11×10^{-6}/°C for steel

different elongation of the two juxtaposed strips. This is the way many thermostats work; as the bimetal strip bends due to temperature changes, it can make or break an electrical contact that operates the heating. A similar bimetal strip is the most commonly used method for temperature compensation in modern variable bypass vaporizers (see Fig. 15.2) [4, 16]; as temperature decreases, the bimetal strip bends so that an increased flow of fresh gas is allowed to enter the vaporizing chamber, while the opposite occurs if temperature increases. Other models have simple metal bellows or rods which expand or contract as temperature changes.

References

1. Bokoch MP, Weston SD. Inhaled anesthetics: delivery systems. In: Gropper MA, Cohen NH, Eriksson LI, Fleisher LA, Leslie K, Wiener-Kronish JP, editors. Miller's anesthesia. 9th ed. Philadelphia: Elsevier; 2019. p. 572–637.
2. Butterworth J, Wasnick J, Mackey DC. The anesthesia workstation. In: Morgan & Mikhail's clinical anesthesiology. 6th ed. New York: McGraw-Hill; 2018. p. 47–80.
3. Aston D, Rivers A, Dharmadasa A. Equipment in anaesthesia and critical care, vol. 85. Scion Publishing Limited; 2014.
4. Boumphrey S, Marshall N. Understanding vaporizers. Contin Educ Anaesth Crit Care Pain. 2011;11(6):199–203.
5. Hameroff SR. Anesthetic action and "quantum consciousness": a match made in olive oil. Anesthesiology. 2018;129(2):228–31.
6. Burdick RK, Villabona-Monsalve JP, Mashour GA, Goodson T 3rd. Modern anesthetic ethers demonstrate quantum interactions with entangled photons. Sci Rep. 2019;9(1):11351.
7. Baldassarre D, Scarpati G, Piazza O. Mechanisms of action of inhaled volatile general anesthetics: unconsciousness at the molecular level. In: Cascella M, editor. General anesthesia research. New York: Humana Press; 2020. p. 109–23.
8. Al-Khalili J. The world according to physics. Princeton University Press: Princeton, NJ; 2020.
9. Hameroff S, Penrose R. Consciousness in the universe: a review of the 'Orch OR' theory. Phys Life Rev. 2014;11:39–78.
10. Al-Khalili J, McFadden J. Life on the edge: the coming of age of quantum biology. London: Weidenfeld & Nicolson Ltd.; 2014.
11. Eger EI 2nd. New inhaled anesthetics. Anesthesiology. 1994;80(4):906–22.
12. Katoh T, Ikeda K. The minimum alveolar concentration (MAC) of sevoflurane in humans. Anesthesiology. 1987;66(3):301–3.
13. Weiskopf RB, Sampson D, Moore MA. The desflurane (Tec 6) vaporizer: design, design considerations and performance evaluation. Br J Anaesth. 1994;72(4):474–9.
14. Graham SG. The desflurane Tec 6 vaporizer. Br J Anaesth. 1994;72(4):470–3.
15. Andrews JJ, Johnston RV Jr. The new Tec6 desflurane vaporizer. Anesth Analg. 1993;76(6):1338–41.
16. Cicman JH, Skibo VF, Yoder JM. Anesthesia systems. Part II: operating principles of fundamental components. J Clin Monit. 1993;9:104–11.

Part VII

Electromagnetic Waves and Optics

Light, Air Pollution, and Pulse Oximetry: The Beer–Lambert Law

16

Contents

Where's Physics

Daily life	Attenuation of sunlight by the atmosphere The colors of the sea The depth at which divers need a flashlight Air pollution and traffic ban
Physics involved	Basic properties of waves Electromagnetic waves and light The electromagnetic spectrum Light absorption (Beer–Lambert law)
Clinical practice	Absorption spectroscopy Pulse oximetry Capnography Inhaled anesthetic analyzers The different colors of arterial and venous blood

A. Pisano, *Physics for Anesthesiologists and Intensivists*,
https://doi.org/10.1007/978-3-030-72047-6_16

16.1 A Journey Through the Waves

This is the first of a series of chapters dealing with *waves* (and surroundings). Here and in Chap. 17 we will discuss about *light* waves (or, more generally, *electromagnetic waves*), while in Chap. 18 we will deal with light from a different point of view (I wanted to write "under a different light," but it would have been a terrible pun!). Finally, in Chaps. 19 and 20 it will be the turn of *oscillations* and *mechanical waves* (such as *sound waves*).

Waves are everywhere in our lives: the colors and the sounds of nature, the different sounds of musical instruments, surfing and earthquakes, most of the technological tools we use every day (including DVDs, Blu-ray discs, remote controllers, smartphones, and microwave ovens, to name but a few), many medical devices (e.g., all the diagnostic imaging, practically), speeding fines, and even the strange way in which spaghetti break (see below) rely on waves and their behavior.

Why talk about waves, light, sound, and oscillations in a book for anesthesiologists and intensivists, though? At least two physical phenomena involving electromagnetic waves, and the laws which describe them, have many important applications (and implications) in daily practice of anesthesia and intensive care medicine and, consistent with the spirit of this book, allow to provide a lot of interesting examples from everyday life. In particular, this chapter addresses the so-called *Beer–Lambert law*, which describes the attenuation of light as it passes through absorbing substances: *pulse oximetry* (and some of its limitations) and certain methods of respiratory gas monitoring (such as *capnography*) are based on it. Other gas analyzers rely, instead, on light *scattering*, which is primarily involved in *near-infrared spectroscopy (NIRS) regional oximetry* (being, moreover, the cause of its main limitation), as well as in some aspects of *radioprotection*, as discussed in Chap. 17. Treating light as a "ray" rather than a wave (see Sect. 16.2) will allow us to understand, in Chap. 18, how the *optical fibers* of our laryngoscopes and bronchoscopes work, while the physics of oscillations and sound waves will take us to the world of *ultrasonography* and, perhaps a little more surprisingly, of invasive pressure monitoring.

Let's start from light and some general features of waves.

16.2 What is Light

The description of light that is closest to our everyday experience is provided by the so-called *ray model*, according to which light travels in straight lines (see Chap. 18). Within certain limits, this model works well. However, a set of "rays" (not interacting with one another) is not what light *really* is. In fact, light undergoes *interference*, which is a typical behavior of waves: the *superposition* of two or more waves produces a resulting wave, or "interference pattern," with a different *amplitude* (see below). Light can be described, therefore, as a wave (in particular, an electromagnetic wave), although it behaves, sometimes, as a (massless) *particle*: the *photon* (according to the so-called *photon model of light*). In reality, in the world of molecules, atoms, and subatomic particles (including the photon!), *all* particles *are* also waves, and vice versa. As mentioned in Chap. 15 (Box 15.1), this is a fundamental point of *quantum physics* (or *quantum mechanics*). As I have already said, this topic is definitely beyond the scope of this book although I don't always resist the temptation to mention it briefly. However, if you want to know more, I suggest you some books that I found particularly enlightening (once again, pardon the pun) [1–4].

Rather we need to know something more about electromagnetic waves.

16.2.1 Light as a Wave

Generally speaking, a wave is a disturbance (i.e., a change in some physical property or dimension) which travels through a "medium," carrying *energy*. Specifically, waves are produced by *oscillations* of such "dimension," i.e., repetitive (or periodic) displacements, both over space and time, around an equilibrium value. Accordingly, they can be described by *sinusoidal* (i.e., sine or cosine) *equations*, or as the sum of sinusoidal waves (see Chap. 19). Waves are usually classified according to the medium in which the cyclic disturbance propagates:

1. *Mechanical waves* need a material medium. Sea waves, for example, travel through water (which is, accordingly, "the medium"), while the oscillating "dimension" is the height of the ruffle. This is almost always the first example that comes to mind when talking about waves and can be found, practically, in all physics textbooks. I would add, therefore, a more intriguing one. Have you ever noticed that spaghetti almost never break in half, but at least in three or four fragments, when bended while holding them at their ends? As suggested in a study from a few years ago, this may be due to *flexural waves* (i.e., changes in the curvature) traveling along the two fragments which are formed after the first breaking [5]. Also sound waves (see Chap. 19) are mechanical waves; in this case, the medium can be air, water, or something else, while what oscillates is density (and, hence, pressure).
2. *Electromagnetic waves* (Fig. 16.1) do not need a material medium and, accordingly, can travel through a vacuum. What oscillates is the *strength* of the *electric field* and of the *magnetic field*, which are perpendicular to each other. We discussed the concept of field in general (in physics), and the electric field in particular, in Chap. 11. As you may remember, the term "field" refers to something which has a defined value at every point in the space; for example, an electric field, which forms around a charged particle *A*, exerts on a second charged particle *B* a *force* (called electrostatic force) which depends on the position of *B* in the space surrounding *A*, and whose direction is always parallel to that of the *electric field vector* (see Sect. 11.1.2) in that point. Similarly, the magnetic field is the force field that is generated in the space surrounding moving charges (e.g., an electric current through a wire) or a permanent magnet; it exerts on any charged particle in that space a *magnetic force* whose direction is, unlike electrostatic

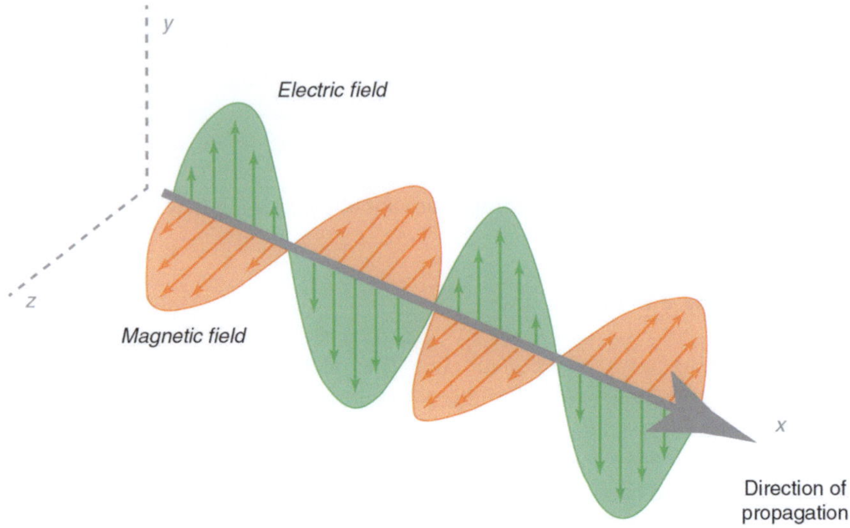

Fig. 16.1 In an electromagnetic wave, the strength of the electric field and that of the magnetic field oscillate perpendicularly to each other and to the direction of propagation

force, always *perpendicular* to that of the magnetic field vector in that point. As shown in Fig. 16.1, electromagnetic waves travel perpendicularly to the planes in which the electric and magnetic fields oscillate; accordingly, they are *transverse waves.*

A third type of waves, called *matter waves*, has to do with the above-mentioned strange affair of particles which, according to quantum physics, behave also as waves (and, of course, does not concern us here).

All waves are characterized by the following properties: *amplitude*, *period, frequency,* and *wavelength* (see Fig. 16.2).

The *amplitude* is the maximum "displacement" produced by the wave (Fig. 16.2a, b). The amplitude of an electromagnetic wave represents its *intensity* (symbolized *I*), that is the mean *energy* transported by the wave over one cycle of oscillation, or the rate at which the wave transfers energy to a certain surface, i.e., *energy per unit area and unit time* (see also Chap. 17).

The *period T* of oscillation of a wave is the time interval in which a complete cycle occurs. It corresponds to the time interval between two wave crests (or between two any repetitions) in a *graph* displaying the "displacement" which occurs at one point in space *over time* (Fig. 16.2a). The frequency *f* is simply the number of oscillations per second. The unit of frequency is s^{-1} or *Hertz* (Hz). Evidently, it is:

$$f = \frac{1}{T} \tag{16.1}$$

Fig. 16.2 Some of the main properties of sinusoidal waves (see text). *t* time; *x* space; *y* "displacement" (change in some physical property or dimension)

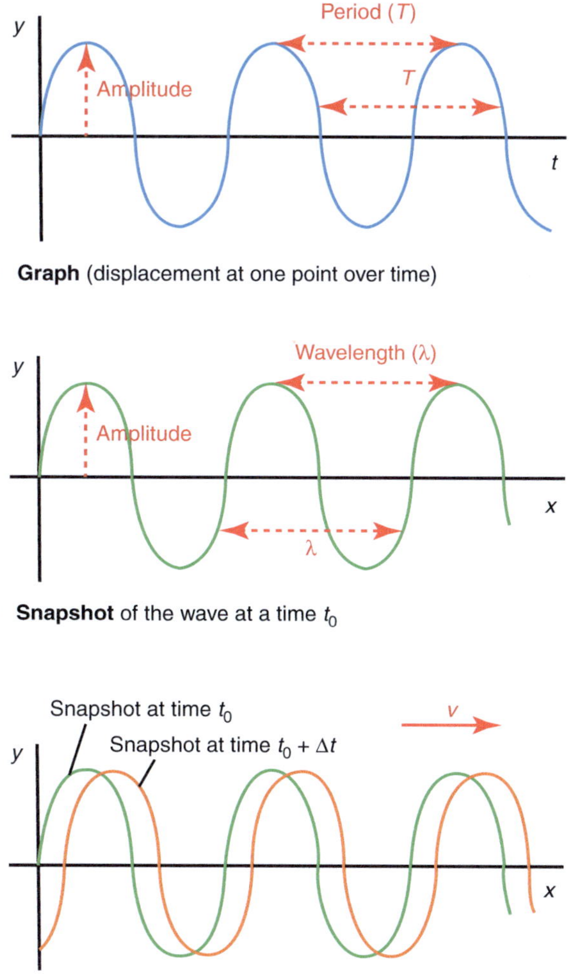

a **Graph** (displacement at one point over time)

b **Snapshot** of the wave at a time t_0

c The wave propagates *through the space* with speed *v*

The *wavelength* λ is the distance between repetitions of the wave shape (e.g., two crests), as can be identified in a *snapshot* of the wave which shows the "displacement" *as a function of distance* (i.e., of the position in space) at a certain time instant (Fig. 16.2b). In other words, the wavelength arises from the propagation of the wave through the space with a certain speed (see Fig. 16.2c) and represents *the distance traveled by the wave in a time interval of one period.*

Remembering that speed is, roughly, distance divided by time (see Chaps. 1, 4, and 7), the *wave speed* v (i.e., the distance traveled by the wave during a time interval Δt) is:

$$v = \frac{\lambda}{T} \tag{16.2}$$

or, substituting for T from Eq. (16.1):

$$v = \lambda f \tag{16.3}$$

Note that the wave speed is a property of the medium in which the wave travels. Since electromagnetic waves don't need a material medium, they have their own speed *in vacuum* (although it decreases in materials such as water or glass), which is the same regardless of the type of electromagnetic wave (see below); this is the *speed of light* (symbolized *c* from Latin "celeritas," i.e., speed), which is about 300,000 kilometers/second (or 186,000 miles/second), as mentioned in Chap. 14. We'll find out more about the speed of light in Chaps. 18 and 21.

For electromagnetic waves traveling in a vacuum, therefore, Eq. (16.3) becomes:

$$\lambda f = c \tag{16.4}$$

that is, *wavelength and frequency are inversely proportional* (since the speed of light *c* is a constant).

16.2.2 The Electromagnetic Spectrum

Visible light is an electromagnetic wave with a wavelength of about 400 to 700 nanometers (nm). As shown in Fig. 16.3, this is only a very small portion of the so-called *electromagnetic spectrum*, i.e., the whole range of the possible wavelengths of electromagnetic waves.

In order of decreasing wavelength (or, according to Eq. (16.4), increasing frequency), the electromagnetic spectrum includes, among others: radio (and TV) waves, microwaves, infrared (IR) radiation, visible light, ultraviolet (UV) radiation, X rays, and gamma rays. Within the range of visible light (400–700 nm), we perceive each wavelength as a different color; for example, the shorter wavelengths (higher frequencies) correspond to blue/violet light, while the longer wavelengths (lower frequencies) are perceived as orange/red.

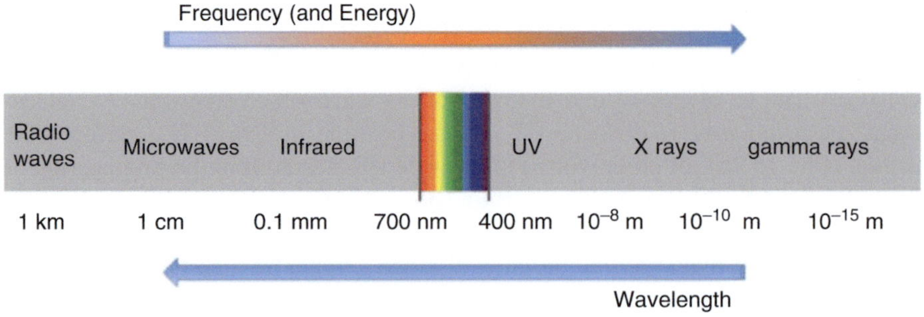

Fig. 16.3 A "qualitative" representation of the electromagnetic spectrum (the distances are not to scale). It can be seen, however, by reading the values of the wavelengths, that visible light is only a very narrow part of the entire range of electromagnetic waves

The higher the frequency (or the shorter the wavelength), the greater the energy of the electromagnetic wave (for a given wave intensity). According to quantum physics, many quantities in the world of atoms and subatomic particles only exist as integer multiples of a minimum (or elementary) amount, called *quantum*. This also applies to electromagnetic waves: the above-mentioned photon represents, in fact, the quantum of light (or, in general, of electromagnetic radiation). For an electromagnetic wave of frequency f, the energy of a single photon (*photon energy*) is:

$$E = h\,f \tag{16.5}$$

where h is a very small quantity called *Planck constant* (see Box 16.1).

Box 16.1: The Planck Constant

The Planck constant (symbolized h) is a fundamental physical constant which jumps out continuously when studying quantum physics (this stunning branch of physics was born at the beginning of the twentieth century precisely from Eq. 16.5). For example, its "reduced" form $\hbar$ (it reads "h-bar"), that is $h/2\pi$, appears in the equation for the Heisenberg's uncertainty principle mentioned in Chap. 15 (Box 15.1), which states (among other things) that we cannot know with high accuracy both the position and the momentum (i.e., the product of mass and speed) of a particle: mathematically, this is expressed by the fact that the product of the "uncertainty" about position and that about momentum (speed) cannot be lower than a very small number, namely $\hbar$. The term "very small" probably doesn't quite convey the idea; to be precise, the value of h is 6.63×10^{-34} J s, i.e., 6.63 divided by ten million billion billion billion! After all, as already said in Chaps. 14 and 15, quantum physics describes the world of very small objects (and $\hbar$ has also to do with the calculation of the smallest length imaginable in nature, the so-called Planck length).

Hence, one photon of violet light has a higher energy than one photon of red light, as well as the energy carried by an X-ray photon is much greater than that transported by a microwave photon (as one can easily guess). The photon energy, however, must not be confused with the energy carried by a certain electromagnetic wave, i.e., with the wave intensity: for a light of a given wavelength, a higher intensity simply means that the wave is made of more photons (and, hence, carries more energy).

16.3 Blue Oceans and Sea Deeps: Beer–Lambert Law

The *Beer–Lambert law*, also known as *Beer's law* (or Lambert–Beer law) correlates the absorption of an electromagnetic wave, as it passes through an object, with the thickness of the object and the amount and absorbing properties of the substances contained therein.

In particular, when a beam of *monochromatic light* (i.e., an electromagnetic radiation of a single wavelength) of intensity I_0 crosses a *solution* by a length L (see Fig. 16.4), the intensity I of the transmitted light decreases according to the following equation:

$$I = I_0 e^{-\varepsilon_\lambda ML} \tag{16.6}$$

where e is the base of natural logarithm, L is the path length, M is the molar concentration (or molarity) of the solution, and ε_λ is the so-called *molar attenuation coefficient* (or *molar extinction coefficient*), which is a characteristic of the solute and depends on the wavelength (in fact, it is here symbolized ε_λ to indicate that it is a function of wavelength). In practice, ε_λ is a measure of how much a chemical species attenuates light at a given wavelength.

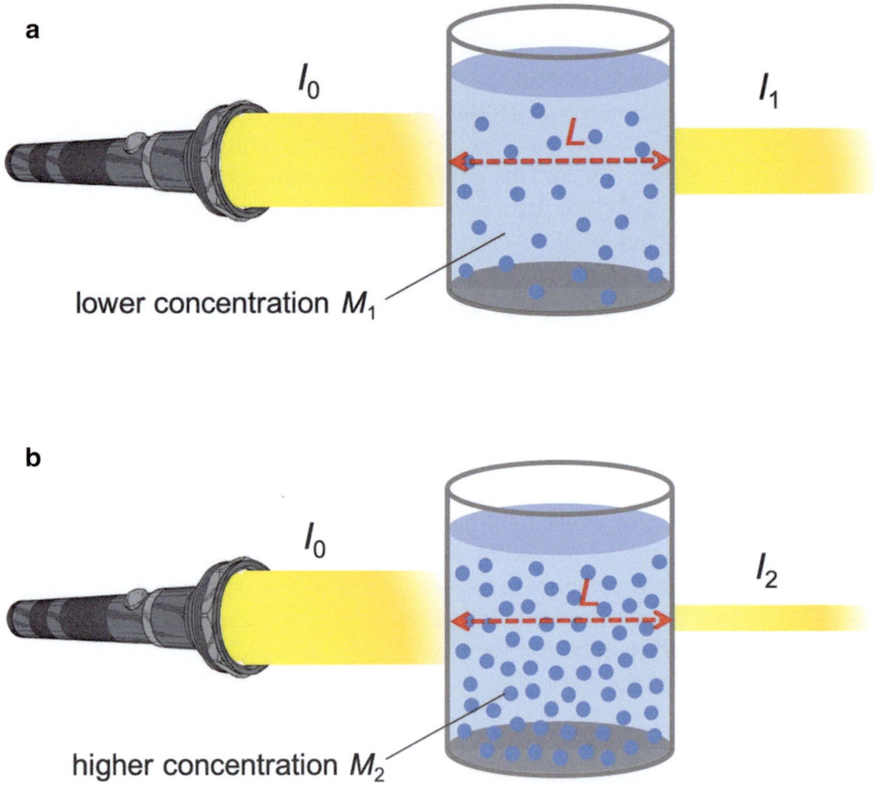

Fig. 16.4 Beer–Lambert law. When a monochromatic light beam of intensity I_0 crosses (by a length L) a solution containing a substance which absorbs light of that wavelength, the intensity of the transmitted light decreases exponentially with the increase in the concentration of the absorbing substance. Since the concentration in **a** is lower than in **b**, it is $I_1 > I_2$. Note that a flashlight like that drawn here does not emit monochromatic light (but, in this way, the figure is prettier!)

If we define the *absorbance A* of the solution as log I_0/I, Eq. (16.6) can be written in the following simpler form:

$$A = \varepsilon_\lambda \, M \, L \tag{16.7}$$

Hence, the higher the concentration of the solution (and the greater the path length), the higher the absorbance, i.e., the lower the intensity of the transmitted light (Fig. 16.4).

A more complex version of this law allows to describe the attenuation of sun radiation as it travels through the atmosphere. Moreover, Beer's law accounts for the color of the sea (due to the dependence of the attenuation coefficient on the wavelength). In fact, the absorbance of seawater is higher for the longer wavelengths (those of red and orange), while shorter wavelengths (such as those of green, blue, and violet) are absorbed to a lesser extent and, accordingly, are more available for *scattering*, i.e., deviation in all directions (see Chap. 17). Since sunlight contains a poor amount of violet light, and our retina is poorly sensitive to violet [6], the sea looks blue (or green in some wonderful places!). Again according to Beer's law, the view around a diver will appear more and more blue as depth increases, until a flashlight will be needed (in fact, the absorbance for any wavelength increases with the path length).

Beer–Lambert law can be used to determine the concentration of a substance in a solution or in a gas mixture, according to the absorption of a (visible, IR, or UV) light of an appropriately chosen wavelength (*absorption spectroscopy*). At its simplest, if I have a solution which contains only one absorbing species (of which I know the molar attenuation coefficient for a given wavelength), I can irradiate a sample (of a certain thickness) of such a solution with a beam of light (of known intensity) of that wavelength and *measure* (thanks to a light detector) the intensity of the transmitted light; according to Eq. (16.6), the only unknown is the concentration of the absorbing species that can be therefore calculated. Some systems for the analysis of polluting substances in the air rely on this technique; also thanks to Beer's law, therefore, we can realize that our cities are polluted (and sometimes the authorities are compelled, for example, to ban the car traffic). What interests us, however, is that many devices we routinely use in the operating room and intensive care unit are also based on this physical law.

16.4 The Beer–Lambert Law in Anesthesia and Critical Care

As mentioned, pulse oximetry, capnography, and some other respiratory gas monitoring systems such as anesthetic (vapor) analyzers rely on absorption spectroscopy and, accordingly, on Beer–Lambert law. The last part of this chapter mainly addresses the physical and technological bases (and some of their clinical implications) of pulse oximetry. Also regional (e.g., cerebral) oximetry is based on Beer's law, but it also involves another physical phenomenon, i.e., the scattering of electromagnetic radiation, and will therefore be discussed in Chap. 17.

16.4.1 Pulse Oximetry

As we all know, pulse oximetry is the noninvasive measurement (or, better, estimation) of *arterial oxygen saturation* (S_aO_2), that is, roughly (i.e., considering the blood content of methemoglobin and carboxyhemoglobin as negligible), the ratio of the concentration of *oxyhemoglobin* (O_2Hb) to the sum of the concentrations of O_2Hb and *reduced hemoglobin* (Hb):

$$S_aO_2 = \frac{[O_2Hb]}{[O_2Hb]+[Hb]} \tag{16.8}$$

Such measurement is possible, by means of absorption spectroscopy, thanks to the different molar attenuation coefficients (ε_λ) of oxyhemoglobin and reduced hemoglobin for most wavelengths (see Fig. 16.5) [7–11].

In particular, a common pulse oximeter (Fig. 16.6) uses two monochromatic *light emitting diodes* (LEDs) to alternately irradiate a finger (or an earlobe) with two wavelengths for which the difference between the molar attenuation coefficients of O_2Hb and Hb is particularly wide: 660 nm (red visible light) and 940 nm (infrared radiation) [7–9, 11]. Even more importantly, as shown in Fig. 16.5, the molar attenuation coefficient of Hb is higher than that of O_2Hb at 660 nm, while the opposite occurs at 940 nm; accordingly, Hb absorbs more red (visible) light, but less infrared radiation, as compared with O_2Hb [7, 9, 11]. This enables the pulse oximeter to "understand" if a variation in light absorption is due to a change in the concentration

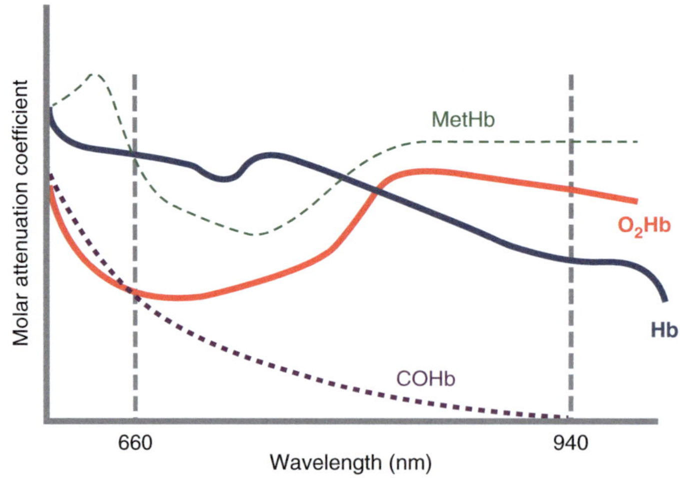

Fig. 16.5 Values of the molar attenuation coefficient of the different forms of hemoglobin as a function of wavelength. Note the differences between oxyhemoglobin (O_2Hb) and reduced hemoglobin (Hb) at the two wavelengths (660 nm and 940 nm) used in common pulse oximeters. The greater absorption of red light (660 nm) by reduced hemoglobin also accounts for the darker appearance of venous blood as compared to the "more red" arterial blood. *MetHb* methemoglobin; *COHb* carboxyhemoglobin

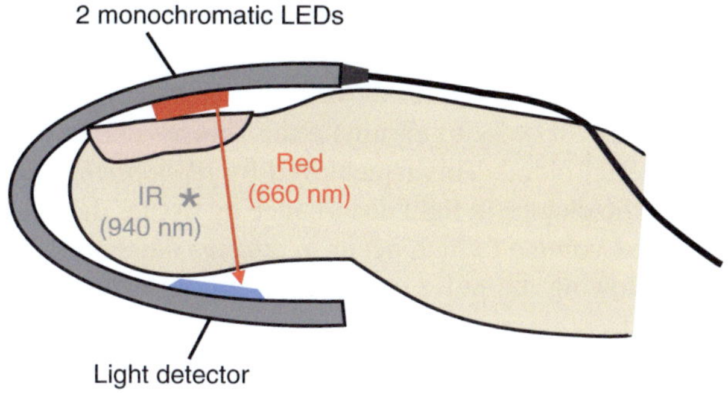

Fig. 16.6 Schematic representation of a pulse oximeter. The asterisk indicates the infrared (IR) beam (which I have not drawn as an arrow because it is not visible!). *LEDs* light-emitting diodes

of reduced hemoglobin relative to that of O_2Hb (i.e., to a change in oxygen saturation) or rather, for example, to a change in the total hemoglobin concentration [9].

In practice, one of the two LEDs emits light (say, at 660 nm) while the other one is off; the absorption of this light is measured (as absorbance) by means of a light detector on the other side, which measures the intensity of the transmitted light (see Fig. 16.6). Subsequently, the absorbance for the second wavelength (940 nm) is measured while the first LED is turned off. The *ratio* between the two absorbances is then measured and compared with those found in healthy volunteers for empiric calibration [7, 10, 11]. If the total hemoglobin concentration decreases (e.g., due to blood loss or hemodilution), the absorbance will be similarly reduced for both wavelengths, and the ratio will not change. Conversely, a reduction in oxygen saturation means that O_2Hb decreases as much as Hb increases (but the total hemoglobin concentration does not change); this leads to an increased absorbance for the 660 nm (red) wavelength (since it is increased, in proportion, the concentration of the species with the higher ε_λ for that wavelength) and to a reduced absorbance for the 940 nm (IR) wavelength (where the situation is reversed). Accordingly, the ratio between the two absorbances (red-to-infrared) increases, and this is translated by the oximeter software into a reduction in saturation.

As you can see, the situation is much more complex than the ideal case, discussed in the previous section, of a solution containing only one species which absorbs light. Since both Hb and O_2Hb, as well as other light absorbers which are present along the path length, absorb light at both wavelengths, in fact, pulse oximeters are not able to measure the absolute concentrations of the two forms of hemoglobin. Accordingly, the red-to-IR absorbance ratio must be matched with the ratios measured in healthy volunteers (in whom S_aO_2 is measured, at the same time, on blood samples) in order to get a value of oxygen saturation; for example, a red-to-IR ratio of 1 corresponds to an S_aO_2 of 85%, while a ratio of 0.43 corresponds to 100% [7]. This is one of the reasons why the accuracy of pulse oximeters significantly decreases for S_aO_2 values below 80% (it is not "nice" to go too down with arterial oxygen saturation in healthy volunteers!) [7, 9].

The accuracy of pulse oximeters may also be affected by high levels of carboxyhemoglobin (COHb) or methemoglobin (MetHb) [7–9, 11]. For example, since the molar attenuation coefficient of COHb at 660 nm is very close to that of O_2Hb (see

Fig. 16.5), pulse oximeters significantly overestimate S_aO_2 values in the presence of non-negligible amounts of COHb.

Of course, the entire above-mentioned process must be only applied to what "pulsates," in order to eliminate the "noise" due to light absorption by tissues and nonarterial blood. This is achieved by *photoplethysmography*, i.e., the analysis of the cyclic change in light absorbance by tissues due to the systo-diastolic variations of blood volume (which occur in arteries much more than in veins and capillaries, thus allowing the pulse oximeter to distinguish arteries from other blood vessels) [11, 12].

Finally, the power cycle of the two LEDs (repeated many times a second) also includes a phase in which both LEDs are switched off, in order to reduce the interference from ambient light. The artifacts related to pulse detection and ambient light, as well as other well-known sources of error, are fully described elsewhere [7–12].

16.4.2 Capnography and Anesthetic Analyzers

Both *mainstream* and *sidestream capnometers* rely on absorption spectroscopy and, accordingly, on Beer–Lambert law [10, 13]. A monochromatic beam of infrared radiation with a wavelength of 4.28 μm, for which carbon dioxide (CO_2) exhibits an absorption peak [14], is used to measure the partial pressure of CO_2 (pCO_2) in the exhaled gas. In mainstream capnometers (nowadays rarely used in the operating room), the absorption detecting device (including both the light emitting element and the light intensity detector) is positioned at the patient end of the respiratory circuit, while in sidestream capnometers a sample of exhaled gas is aspirated within the measurement device through a small-bore tube. The latter type of capnometers is often part of (sidestream) infrared absorption spectroscopy *respiratory gas analyzers*, which are usually integrated in the anesthesia workstation. These devices use different wavelengths appropriately chosen to measure, in addition to pCO_2, the concentration of other respiratory gases such as nitrous oxide and different volatile anesthetics (anesthetic analyzers).

References

1. Feynman RP. Six easy pieces. New York: Basic Books; 2011.
2. Quantum A-KJ. A guide for the perplexed. London: Weidenfeld & Nicolson; 2004.
3. Lederman LM, Hill CT. Quantum physics for poets. New York: Prometheus Books; 2011.
4. Al-Khalili J. The world according to physics. Princeton, NJ: Princeton University Press; 2020.
5. Audoly B, Neukirch S. Fragmentation of rods by cascading cracks: why spaghetti do not break in half. Phys Rev Lett. 2005;95(9):095505.
6. Walker J. The flying circus of physics. 2nd ed. New York: Wiley; 2007. p. 245.
7. Jubran A. Pulse oximetry. Crit Care. 1999;3(2):R11–7.
8. Jubran A. Pulse oximetry. Crit Care. 2015;19:272.

9. Aston D, Rivers A, Dharmadasa A. Equipment in Anaesthesia and critical care. Branbury: Scion Publishing; 2014. p. 189–91.
10. Szocik JF, Barker SJ, Tremper KK. Fundamental principles of monitoring instrumentation. In: Miller RD, editor. Miller's anesthesia. 6th ed. Philadelphia, PA: Elsevier Churchill Livingstone; 2005. p. 1191–225.
11. Pisano A, Di Fraja D, Angelone M. Pulse oximetry and noninvasive ventilation applications. In: Esquinas AM, editor. Applied physiology in noninvasive mechanical ventilation: key practical insights. New Haven: PMPH USA; 2021. p. 157–63.
12. Chan ED, Chan MM, Chan MM. Pulse oximetry: understanding its basic principles facilitates appreciation of its limitations. Respir Med. 2013;107(6):789–99.
13. Langton JA, Hutton A. Respiratory gas analysis. Contin Educ Anaesth Crit Care Pain. 2009;9(1):19–23.
14. Aston D, Rivers A, Dharmadasa A. Equipment in anaesthesia and critical care. Branbury: Scion Publishing; 2014. p. 174–5.

Scattering of Electromagnetic Waves: Blue Skies, Cerebral Oximetry, and Some Reassurance About X-Rays

17

Contents

Where's Physics

Daily life	Why the sky is blue Wet sand vs. dry sand How far car headlights illuminate
Physics involved	Scattering of electromagnetic waves Compton scattering The inverse square law (of light intensity) Raman scattering Light absorption and Beer–Lambert law (see Chap. 16)
Clinical practice	Near-infrared spectroscopy (NIRS) cerebral oximetry A few elements of radioprotection Some types of respiratory gas analyzers Why the veins look blue

A. Pisano, *Physics for Anesthesiologists and Intensivists*,
https://doi.org/10.1007/978-3-030-72047-6_17

17.1 Electromagnetic Waves Encounter Matter: Scattering

The trajectory of an electromagnetic wave which travels through a (substantially) transparent medium may be deviated in any direction as a result of the *interaction* between the wave and some obstacle or "inhomogeneity," such as the particles (including atoms or molecules) that are suspended in the medium. This phenomenon, which is referred to as *scattering* of the electromagnetic wave, should not be imagined as a "mechanical collision" with a body on which the wave "bounces." What happens indeed is that the oscillating electric field of the electromagnetic wave (see Chap. 16) causes the electrons within the atoms or molecules of which the body is made to move "up and down" [1]. As a consequence, the negative charges (i.e., the center of the electron cloud) *periodically* approach and move away from the positive charges (i.e., the center of the nucleus), producing what is called an oscillating *induced dipole moment* (see Box 17.1 if you are curious, otherwise read on). In practice, the atoms or molecules of the "obstacle" which is encountered by the electromagnetic radiation acquire part of the energy of the incident wave and become themselves oscillating electric fields, which act as a source of "*new*" electromagnetic radiation that propagates in different directions, always perpendicular to the direction of vibration of the charges (since, as mentioned in Chap. 16, electromagnetic waves are transverse waves, i.e., they travel perpendicularly to the direction in which the oscillations of the electric field occur, that is the same in which the electrons vibrate) [1].

Box 17.1: Induced Dipole Moment

Two equal and opposite (i.e., positive and negative) electric charges, separated by a small distance, form an *electric dipole*. Many molecules, such as that of water, have a permanent dipole, i.e., positive and negative charges maintain a slight separation with one another. However, a neutral atom can be *polarized* by means of an external electric field (see Chap. 11), in order to obtain an *induced dipole*; since the electrons are located peripherally, the external electric field attracts them a bit more than it repulses the nucleus, thus leading to a small separation of charges. An electric dipole, both permanent or induced, has an electric field whose strength is determined by the magnitude of the so-called *dipole moment* p:

$$p = q d$$

where $\pm q$ is the value of the positive and negative electric charges and d is the distance between them. The dipole moment is a vector (see Chap. 13) which points from the negative to the positive charge; accordingly, it describes both the orientation and the strength of the dipole's electric field.

If an electric field induces a dipole moment within the context of an atom, an *oscillating* electric field like that of an electromagnetic wave will produce an *oscillating induced dipole moment*, i.e., a periodical widening and narrowing of the distance between the positive and negative charges of the atom and, accordingly, a second oscillating electromagnetic field associated to the oscillating dipole moment. That's how an incident electromagnetic wave produces a scattering wave.

For some aspects, however, the scattering of electromagnetic waves can be roughly likened to the collisions which occur between two (macroscopic) bodies. According to classical mechanics, a collision is said to be "elastic" if the total kinetic energy (see Chap. 2, Box 2.2) of the system made up of the two colliding objects is conserved, while is called "inelastic" if the total kinetic energy of the system changes after the collision. For example, the collisions of the molecules of a gas against the walls of its container (see Chap. 2) are essentially elastic; those between two billiard balls are something quite similar to an elastic collision (although they are not really elastic; in fact, not *all* the kinetic energy of the hitting ball is transferred to the one being hit; a small part is converted, for example, into noise); finally, a car crash is definitely an inelastic collision.

Similarly, in *elastic scattering* the photon energy (see Chap. 16) doesn't change; according to Eq. (16.5), therefore, the scattered wave has the same frequency and, hence, wavelength (see Eq. 16.4) of the incident wave. Conversely, *inelastic scattering* involves a change in the energy and, accordingly, in the wavelength of the radiation.

As discussed below, both types of scattering of electromagnetic radiation may have a role in the daily practice of anesthesia and intensive care medicine. For example, the elastic scattering of infrared radiation is one of the principles on which *cerebral oximetry* is based and, at the same time, represents a major technological challenge which accounts for some limitations of this technique. The special kind of (inelastic) scattering which ionizing radiations undergo (the so-called *Compton scattering*) must be considered in order to keep an appropriate distance, for example, during a chest X-ray in the intensive care unit (ICU). Moreover, although non-ionizing electromagnetic waves mainly undergo elastic scattering, inelastic scattering may also occur; this phenomenon, which is known as *Raman scattering*, is used in some types of respiratory gas analyzers. Finally, as we will see at the end of this chapter, light scattering (together with the Beer–Lambert law discussed in Chap. 16) also explains why veins look blue when viewed through the skin.

17.2 Electromagnetic Scattering, Cerebral Oximetry, and Why the Sky Is Blue

As mentioned, non-ionizing electromagnetic waves such as visible light or infrared radiation mainly undergo elastic scattering. Hence, if we imagine a photon of a given wavelength (say, red visible light) which "strikes" an atom, as shown in Fig. 17.1a, such a photon may be diverted into a *random* direction, or better it has a

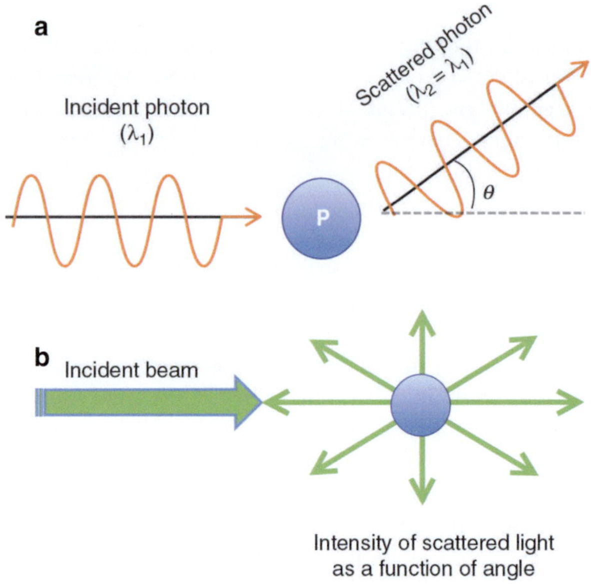

Fig. 17.1 Rayleigh scattering. (**a**) A photon (of non-ionizing radiation) scattered at a certain angle θ from a small particle (P) such as an atom or a molecule has usually the same wavelength of the incident photon (elastic scattering). (**b**) A beam of light (i.e., a huge number of photons) passing through a "scattering medium" is scattered in all directions according to an angular distribution of intensities. Note that the intensity at 90° is one half the backward and forward intensities (see text). λ_1 wavelength of the incident photon; λ_2 wavelength of the scattered photon

certain *probability* to be deviated by a certain angle, but it will maintain the same wavelength (i.e., exactly the same color). Practically, the electrons of the atom will oscillate with the same frequency of the incident light (see Chap. 16).

In daily life, however, we don't deal with one photon and one atom, but usually with a "beam" of light of a certain *intensity*, i.e., consisting of a certain amount of photons, which passes through a body containing many *scatterers* (i.e., particles able to deviate the electromagnetic waves). Accordingly, multiple scattering occurs and light is diverted in *all directions*, with different intensity depending on the scattering angle (Fig. 17.1b). In fact, since the intensity of light can be regarded as the "number of photons" (as discussed in Chap. 16), the intensity of scattered light will be higher in those directions along which the photons have a higher probability to be deviated. In other words, the "probability distribution" of a single scattering event translates into an "angular distribution" in the case of multiple scattering.

Now let's see why the sky is blue (or, better, why it *looks* blue) [1, 2].

The elastic scattering of visible light by tiny particles such as atoms and molecules is described by the so-called Rayleigh theory for electromagnetic scattering (after lord J.W. Rayleigh). This model applies, in general, to the scattering of electromagnetic waves by particles whose diameter is lower than one tenth of their wavelength. In the case of *Rayleigh scattering* by *molecules*, in particular, the intensity of scattered radiation I_{sctd} at any angle θ can be calculated as follows:

$$I_{\text{sctd}} = I_0 \frac{8\pi^4 N\alpha^2}{\lambda^4 R^2}\left(1+\cos^2\theta\right) \tag{17.1}$$

where I_0 is the intensity of the incident light; N the number of molecules (scatterers); α their *polarizability* (which is the tendency of the molecules to be polarized, i.e., as discussed in Box 17.1, the tendency of their opposite charges to be separated from each other by means of an external electric field; accordingly, it expresses how easily their electron cloud vibrates under the effect of the oscillating electric field of an electromagnetic wave); R the distance from scatterers; and λ the wavelength.

Note that, according to Eq. (17.1), the intensity of Rayleigh scattering at 90° is one half of the forward and the backward intensities (see also Fig. 17.1); in fact, remembering a little about trigonometry (see Chap. 1), the coefficient $(1 + \cos^2\theta)$ is equal to 2 for both $\theta = 0°$ and $\theta = 180°$ ($\cos 0° = 1$, $\cos 180° = -1$) and to 1 for $\theta = 90°$ ($\cos 90° = 0$).

However, we have to focus on just one aspect of Eq. (17.1); the intensity of scattered radiation depends inversely on the fourth power of the wavelength:

$$I_{\text{sctd}} \propto \frac{1}{\lambda^4} \tag{17.2}$$

Therefore, the electromagnetic waves with a shorter wavelength are much more scattered than those with a longer wavelength. As sunlight (which contains practically all wavelengths of visible light and is, hence, substantially white) passes through the atmosphere, it is scattered by the air molecules (mainly nitrogen and oxygen). However, according to the Rayleigh's model, the blue and violet components (wavelengths around 400–450 nm) are scattered to a greater extent as compared with red-orange light (650–700 nm). Moreover, as mentioned in Chap. 16, the sunlight does not contain large amounts of violet light and our retina is poorly sensitive to violet. Accordingly, when you look at the sky (in all directions except that of the sun), you see mainly blue light.

Light scattering also explains why wet sand is much darker than dry sand [3]. In fact, light is scattered more times in wet sand than in dry sand. For example, it may be entrapped into a water layer, where it is absorbed progressively as it is repeatedly scattered. Moreover, when the sand grains are wet (and, hence, bigger), light is more likely to be scattered in the forward direction, i.e., deeper in the sand (according to the Mie theory of scattering, which usually applies when the scatterers are "particles" larger than molecules). In both cases, a lesser amount of light "comes out" from the sand, which looks darker accordingly.

It's now time to see what all this has to do with cerebral oximetry.

17.2.1 The Unknown of Cerebral Near-Infrared Spectroscopy

In the last two decades, the monitoring of regional (especially cerebral) oxygen saturation of hemoglobin by means of *near-infrared spectroscopy* (NIRS) has been increasingly used, as a measure of oxygen supply/consumption balance, in several clinical settings [4], including cardiovascular surgery [5–12], pediatric anesthesia and ICU [13–15], interventional neuroradiology procedures [16], and cardiac arrest [9, 17–19]. In particular, some evidences suggest that the use of intraoperative cerebral oximetry, as well as the implementation of strategies aimed at preventing clinically significant cerebral desaturations, may help to improve important outcomes in patients undergoing cardiac surgery [9, 11, 20–23]. Moreover, a multicenter investigation showed an association between higher cerebral oximetry values during cardiopulmonary resuscitation (CPR) and both the return of spontaneous circulation (ROSC) and a more favorable neurological outcome after cardiac arrest [18], and a more recent study also showed an increased chance of ROSC in patients whose cerebral oximetry values increased by at least 15% during CPR as compared with the initial values measured [19]. However, there are still not sufficient data which support the routine use of NIRS oximetry, whose clinical usefulness in anesthesia and critical care medicine is indeed hotly debated [4, 7–9, 13, 24, 25]. This is also due, maybe, to some limitations of this technique, including the lack of a "reference" value and of reliable absolute measurements [4, 8, 26], the poor agreement among different devices [4, 10, 27, 28] and, accordingly, the lack of definite thresholds for either "normal values" or clinically important desaturations [4, 9, 10, 24].

As discussed below, the scattering of electromagnetic radiation, on which NIRS partly relies, is also largely responsible for these limitations.

Like a pulse oximeter, an NIRS cerebral oximeter uses the Beer–Lambert law to estimate oxygen saturation thanks to the different molar attenuation coefficients of oxyhemoglobin and reduced hemoglobin for appropriately chosen wavelengths (see Chap. 16). In common commercially available NIRS devices, an electromagnetic radiation of two or more different wavelengths within the near-infrared (NIR) range (700–950 nm, i.e., the part of the infrared spectrum closest to the visible wavelengths) is emitted by as many light emitting diodes (LEDs) or other types of emitting "optodes" [4]. The intensity of the radiation that, after having crossed a small area of the brain, reaches a couple of detecting optodes (positioned at a certain distance from one another) is then measured. NIR wavelengths have the advantage to be preferentially absorbed, with distinct absorption spectra, by a few substances (including hemoglobins), while the largest part of brain and other tissues is relatively transparent to them [8]. Nevertheless, since our head is much bigger than a finger, the emitting and detecting optodes cannot lie on opposite sides of the brain (as in the case of a finger in the pulse oximeter) because no radiation would reach the light detector due to the excessive path length [9].

However, infrared radiation undergoes scattering as it passes across brain; as in the case of sunlight passing through sand, some radiation comes back (or, as it is said, is "back-scattered") after a certain number of scattering events, thus reaching the detecting optodes which are located a few centimeters away from the emitting optodes on a single probe, usually an adhesive patch which is placed on the forehead (Fig. 17.2). Two patches are usually applied in adults (one on the left and the other on the right).

Hence, cerebral oximetry is only possible thanks to scattering. Unfortunately, it's here that problems begin! In particular, the issues are two:

1. Unlike pulse oximetry, the exact path length of the electromagnetic radiation, which you need to know in order to calculate the absorbance according to Beer–Lambert law (see Chap. 16), is unknown. For example, we don't know precisely how many times the infrared radiation has been scattered before coming out from the brain and reaching the detecting optode (see Fig. 17.2).
2. The electromagnetic radiation that reaches the detecting optodes has not been attenuated only (or mostly) due to absorption, as in the case of pulse oximetry, but mainly due to scattering. In other words, *regardless of absorption by hemoglobins*, the intensity of the electromagnetic wave measured by the detecting optodes will be reduced because most of the radiation takes other directions (just take another quick look at Eq. (17.1) and Fig. 17.1).

Accordingly, in the case of NIRS oximetry Eq. (16.7) (Beer–Lambert law) assumes the following form [8, 29]:

$$A = \log \frac{I_0}{I} = \varepsilon_\lambda \, M \, L \, DPF + G \tag{17.3}$$

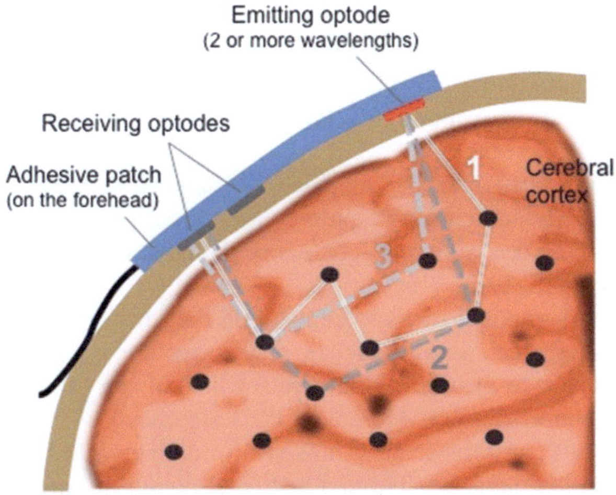

Fig. 17.2 Schematic representation of an NIRS cerebral oximetry probe. The simplified illustration shows that an infrared "ray" undergoes an unknown number of scattering events and, accordingly, may have followed different "true" path lengths (such as number 1, 2, or 3) before reaching the detectors

where A, I_0, I, ε_λ, M, and L are, respectively, the absorbance, the intensities of provided and measured radiation, the molar attenuation coefficient, the concentration (molarity) of the absorbing substance, and the distance between the emitting and the detecting optodes (as in Eqs. 16.6 and 16.7), while *DPF* is the so-called *differential pathlength factor*, a coefficient of proportionality which accounts for the actual path length, and *G* is the attenuation due to scattering.

In practice, the actual path length of NIR radiation (i.e., the real optical distance between the emitting and the detecting optodes), which is referred to as the *differential pathlength* (*DP*), is the product of the geometrical distance *L* between the emitting and the detecting optodes and the differential pathlength factor *DPF*:

$$DP = L \times DPF \tag{17.4}$$

The scattering loss *G* depends on both the geometry and the scattering properties of the tissue and is accordingly unknown [29]. However, once the probe is placed on the patient's forehead, it can be assumed that *G* (as well as the *DPF*) remains constant, so that the changes in the measured attenuation are only due to changes in the *absorption* by absorbing species such as hemoglobins; accordingly, the changes in the concentration of oxyhemoglobin and reduced hemoglobin (and, hence, the *changes* in cerebral oxygen saturation) can be measured by using a differential equation between the two species, i.e., measuring the difference in the attenuation at (a minimum of) two different wavelengths [8, 29]. In this way, the "factor" *G* is (mathematically) eliminated.

In order to provide *absolute* values of oxygen saturation (and not only a *trend*), an estimation of the *DPF* is needed. Most commercially available NIRS devices use a "continuous wave" technology and rely on *spatially resolved* (e.g., *multidistance*) *spectroscopy* to estimate the *DPF* [4]. Briefly, the attenuation of infrared radiation (of different wavelengths) is measured by at least two detectors which are located at a different distance from the emitting optodes; in order to estimate the *DPF*, the increase in attenuation with the source-detector distance is assessed and matched with the dependency of scattering on the wavelength (which, as discussed, also explains why the sky is blue). Other NIRS devices use non-continuous technologies such as radio-controlled intensity modulation (*frequency resolved* or *frequency domain* spectroscopy) or ultrashort pulsed laser (*time-resolved* or *time-of-flight* spectroscopy), whose description in details is beyond the scope of this book and can be found elsewhere [8, 29].

The several "assumptions" which are made in the (different) algorithms of the different NIRS oximeters, as well as their relative technological complexity, may partly explain the above-mentioned limitations which hinder the widespread adoption of these monitoring tools. Moreover, the "accuracy" of NIRS oximeters is something very difficult to evaluate. Unlike pulse oximetry, in fact, the measurement is not made into an artery, but in a restricted region of brain which contains veins, arteries, and capillaries in a widely variable ratio (while the proprietary algorithms usually assume a fixed ratio of 25:75 or 30:70 between arteries and veins) [4, 27, 30]. It is evident, though, that the expected value (that the device attempts to measure) does not exist "physiologically," but "is decided" at the same moment in which the probe is placed on the forehead of the patient [4, 26]. Accordingly, there is not a "true" value that can be measured by a "gold standard" technique.

17.3 Catch Me If You Can: X-Rays, Compton Scattering, and the Inverse Square Law

When a patient has a chest X-ray in the ICU, usually all staff flees as far away as possible for fear of being "hit" by radiation. In fact, although the X-ray beam is directed only towards the patient, some radiation undergoes scattering as it interacts with the patient's body. However, as discussed below, you don't need a huge distance to be at safe.

17.3.1 X-Rays ... from a Different Angle: Compton Scattering

As mentioned, unlike visible light or infrared radiation, X-rays (as well as gamma rays) undergo *inelastic scattering* when they encounter a material object. That is, the wavelength of scattered radiation is *longer* than that of the incident radiation (Fig. 17.3). This phenomenon, which is referred to as *Compton scattering* (or *Compton effect*), can be explained by interpreting the interaction between X-ray photons and electrons (in particular, the weakly bound electrons of the atoms within the object) as an *elastic collision*, similar to that between two billiard balls. In fact, the discovery of Compton effect was one of the best proofs of the quantum (or "photon") model of electromagnetic radiation (see Chap. 16), i.e., of the fact that light is a wave that sometimes behaves as a particle (and, accordingly, it can collide with another particle approximately like two billiard balls do with one another). Actually, an increase in the wavelength of scattered radiation occurs for all wavelengths although it is negligible for longer wavelengths, while it becomes significant for the shorter ones (X-rays and gamma rays). In practice, as shown in Fig. 17.3, when an X-ray (or a gamma ray) photon is scattered at a certain angle θ from an electron, also the electron is "sent away" (at an angle ϕ), i.e., its kinetic energy increases.

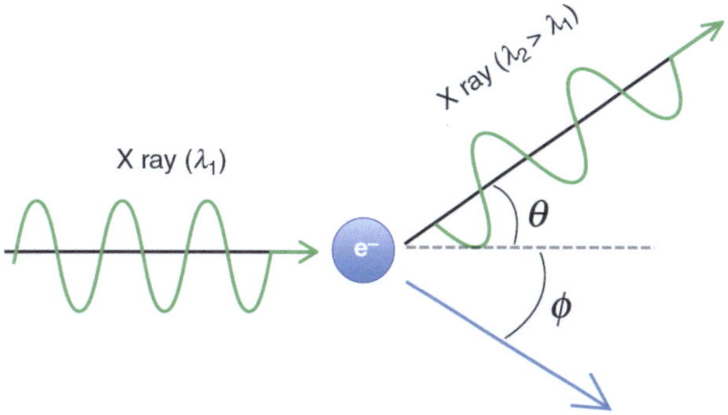

Fig. 17.3 Compton scattering. An X-ray photon of wavelength λ_1 which collides with an electron (e^-) is scattered at an angle θ with *increased* wavelength λ_2, while the electron is "sent" at an angle ϕ

Since the total energy is conserved in an elastic collision, the energy of the photon must decrease. As discussed in Chap. 16, a lower photon energy means a lower frequency (Eq. 16.5) and, according to Eq. (16.4), a longer wavelength.

The *wavelength shift* $\Delta\lambda$ depends on the scattering angle θ according to the equation:

$$\Delta\lambda = \frac{h}{mc}\left(1-\cos\theta\right) \tag{17.5}$$

where h is the Planck constant (which has been already mentioned in Chap. 16), m the mass of the particle from which scattering occurs (e.g., the electron), c is the speed of light, and the coefficient h/mc is known as *Compton wavelength.*

The maximum increase in wavelength (and, according to Eq. (16.5), the maximum decrease in photon energy) of scattered X-ray radiation occurs at 180°, while there is no wavelength shift for X-rays scattered forward ($\theta = 0°$). In fact (see Chap. 1), $\cos 180° = -1$, so $1 - \cos\theta = 1 - (-1) = 2$, and $\cos 0° = 1$, that is, $1 - \cos\theta = 1–1 = 0$. In other words, the energy of scattered photons decreases as the scattering angle increases; accordingly, back-scattered X-ray photons are those with the lowest energy, while X-ray photons which are scattered at 90° have an intermediate energy as compared with those scattered backward and forward ($\cos 90° = 0$, so $1 - \cos\theta = 1–0 = 1$).

Of course, the number of photons which are scattered at a certain angle (as a result of the interaction with the patient) is only a small fraction of the amount of photons contained in the so-called *primary X-ray beam* emitted (only towards the patient) by the X-ray machine. In practice, the scattered radiation at any angle will have a much lower *intensity* (which is, as mentioned in Chap. 16, the *energy* carried by an electromagnetic wave) than that of the incident radiation for two reasons: (a) it consists of a much lower number of photons, (b) each of which has a lower energy.

On the basis of these considerations, it is usually estimated that the intensity of X-rays at an angle of 90° to the primary X-ray beam, *at a distance of 1 m*, is about 1/1000 that of the primary beam [31]. In other words, the *absorbed dose* (expressed in joule/kg or *gray*, symbolized *Gy*, that is the energy of ionizing radiation absorbed by a unit mass of tissue) by someone who lies in this position is approximately 1/1000 that absorbed by the patient. And what about, say, 2 m?

17.3.2 Far Enough Away: The Inverse Square Law

As discussed in Chap. 16, the intensity I of an electromagnetic wave is the energy E per unit time t (which is referred to as *the electromagnetic wave power* $P = E/t$) which is transferred by the wave per unit area A (measured on a plane perpendicular to the direction of propagation of the wave):

$$I = \frac{E}{tA} = \frac{P}{A} \tag{17.6}$$

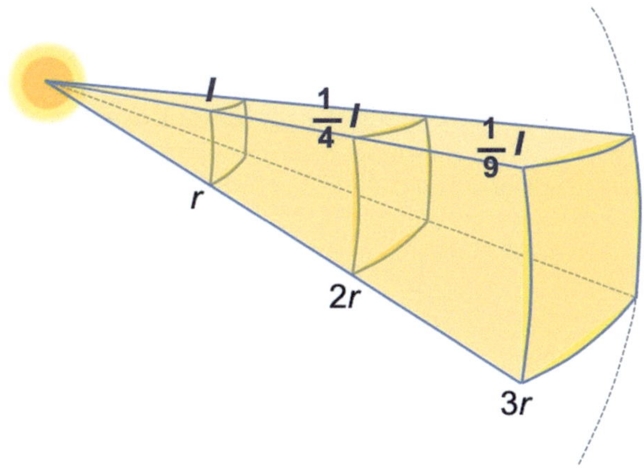

Fig. 17.4 The intensity I of an electromagnetic radiation decreases with the inverse square of the distance r from its (point) source. For example, the headlights of a car which is two times more distant than another from an observer will appear four times fainter. As shown, this is due to the larger surface (that increases with the square of the distance) over which the wave energy is distributed

Considering a *point source* of electromagnetic radiation with power P_s which propagates uniformly (i.e., with the same intensity) in all directions, it can be said that, at a distance r from such a source, the power of the electromagnetic wave is distributed over the surface area of a sphere of radius r, that is $4\pi r^2$ (see Chap. 1 and Box 6.2). Accordingly, the intensity I of the electromagnetic wave at the distance r is:

$$I = \frac{P_s}{4\pi r^2} \tag{17.7}$$

Hence, the intensity of an electromagnetic radiation generated by a point source decreases with the inverse square of the distance from the source (*inverse square law*). This simply comes from the fact that, as the distance from the source increases, the energy carried by the wave is distributed over a larger surface area; evidently, this area increases with the square of the distance (Fig. 17.4). The inverse square law applies, with good approximation, to X-rays scattered from the patient as well as, for example, to the headlights of your car (although they are not exactly point sources). Accordingly, if the intensity of scattered X-rays at 90° is 1/1000 that of the primary beam at a distance of 1 m from the patient, it will be roughly 1/4000 at 2 m, 1/9000 at 3 m, 1/16000 at 4 m, and so on. Hence, you don't need to go too far (unless you take the opportunity for a short break!).

17.4 Electromagnetic Scattering and Gas Analyzers: Raman Spectroscopy

Although, as discussed in Sect. 17.2, non-ionizing electromagnetic waves such as visible light, infrared (IR) radiation, and ultraviolet (UV) radiation mainly undergo elastic scattering from molecules, about one every 10 million scattering events may be *inelastic*, i.e., involve a change in the wavelength of the scattered radiation. This phenomenon is known as *Raman scattering* or "Raman effect" (so named after the

Indian physicist C.V. Raman). Since the wavelength shift depends on the molecular species from which scattering occurs, the concentration of a specific component within a gas mixture can be measured according to the intensity of the scattered radiation of a certain wavelength (*Raman spectroscopy*). Some respiratory gas analyzers used in anesthesia rely on this technique to measure the concentration of gases such as oxygen, carbon dioxide, and nitrous oxide [32–34]. However, due to the extreme rarity of the phenomenon, a very high-intensity radiation source (an argon laser) is needed in order to generate a detectable "Raman signal." For this reason, these devices are much more expensive than those based on infrared absorption (see Chap. 16) and, accordingly, are rarely used in clinical practice.

17.5 Scattering and Absorption of Light Which Crosses the Skin: Why Veins Look Blue

As mentioned, light scattering (see Sects. 17.1 and 17.2) and light absorption (as described by the Beer–Lambert law discussed in Chap. 16) also explain why the veins that we can see through the skin look *blue* although the blood they contain is incontrovertibly red [35]. In fact, due to the wavelength-dependency of both light absorption and scattering by the "chromophores" and "scatterers" which are contained in skin, *red light has a deeper skin penetration than blue light* [36]. Think, for example, only of scattering. According to Eq. 17.1 (or, if you prefer, to Eq. 17.2), the intensity of scattered light is inversely proportional to the fourth power of wavelength. As mentioned in Chap. 16, blue light has a shorter wavelength than red light (see Fig. 16.3); hence, it is more back-scattered and, accordingly, it comes less deep through skin than red light. Let's now compare two different skin areas, one surmounting a vein and one with no veins below. In the first area, blue light is heavily back-scattered before reaching the vein, while more red light continues its journey to the vein, where it is largely absorbed by hemoglobin (see Chap. 16 and, in particular, Fig. 16.5). Also in the second area red light reaches deeper, but it doesn't find a vein filled with hemoglobin, which would absorb much of it; accordingly, red light has more chances to be back-scattered (like blue light) in this area. Unconsciously, we compare these two areas and, since we see *less* red light coming from that with the vein below, we attribute the blue color to the vein (although the same "amount" of blue light comes from the two areas actually) [35].

References

1. Feynman RP, Leighton RB, Sands M. The Feynman lectures on physics: the new millennium edition, vol. 1. New York: Basic Books; 2011. p. 26–32.
2. Walker J. The flying circus of physics. 2nd ed. New York: Wiley; 2007. p. 244–5.
3. Walker J. The flying circus of physics. 2nd ed. New York: Wiley; 2007. p. 261–2.
4. Pisano A, Di Fraja D, Palmieri C. Monitoring cerebral oximetry by near-infrared spectroscopy (NIRS) in anesthesia and critical care: Progress and perspectives. In: Cascella M, editor. General anesthesia research. New York: Humana Press; 2020. p. 75–96.

5. Subramanian B, Nyman C, Fritock M, Klinger RY, Sniecinski R, Roman P, Huffmyer J, Parish M, Yenokyan G, Hogue CW. A multicenter pilot study assessing regional cerebral oxygen desaturation frequency during cardiopulmonary bypass and responsiveness to an intervention algorithm. Anesth Analg. 2016;122:1786–93.
6. Grocott HP. Cerebral oximetry monitoring. To guide physiology, avert catastrophe or both? Eur J Anaesthesiol. 2019;36(1):82–3.
7. Zheng F, Sheinberg R, Yee MS, Ono M, Zheng Y, Hogue CW. Cerebral near-infrared spectroscopy monitoring and neurologic outcomes in adult cardiac surgery patients: a systematic review. Anesth Analg. 2013;116:663–76.
8. Ghosh A, Elwell C, Smith M. Cerebral near-infrared spectroscopy in adults: a work in progress. Anesth Analg. 2012 Dec;115(6):1373–83.
9. Green DW, Kunst G. Cerebral oximetry and its role in adult cardiac, non-cardiac surgery and resuscitation from cardiac arrest. Anaesthesia. 2017;72(Suppl 1):48–57.
10. Pisano A, Galdieri N, Iovino TP, Angelone M, Corcione A. Direct comparison between cerebral oximetry by INVOS(TM) and EQUANOX(TM) during cardiac surgery: a pilot study. Heart Lung Vessel. 2014;6(3):197–203.
11. Deschamps A, Hall R, Grocott H, Mazer CD, Choi PT, Turgeon AF, et al. Canadian perioperative anesthesia clinical trials group. cerebral oximetry monitoring to maintain normal cerebral oxygen saturation during high-risk cardiac surgery: a randomized controlled feasibility trial. Anesthesiology. 2016;124(4):826–36.
12. Maldonado Y, Singh S, Taylor MA. Cerebral near-infrared spectroscopy in perioperative management of left ventricular assist device and extracorporeal membrane oxygenation patients. Curr Opin Anaesthesiol. 2014;27(1):81–8.
13. Kasman N, Brady K. Cerebral oximetry for pediatric anesthesia: why do intelligent clinicians disagree? Paediatr Anaesth. 2011;21(5):473–8.
14. Ghanayem NS, Hoffman GM. Near infrared spectroscopy as a hemodynamic monitor in critical illness. Pediatr Crit Care Med. 2016;17(8 Suppl 1):S201–6.
15. Olbrecht VA, Skowno J, Marchesini V, Ding L, Jiang Y, Ward CG, Yu G, Liu H, Schurink B, Vutskits L, de Graaff JC, McGowan FX Jr, von Ungern-Sternberg BS, Kurth CS, Davidson A. An international, multicenter, observational study of cerebral oxygenation during infant and neonatal anesthesia. Anesthesiology. 2018;128(1):85–96.
16. Badenes R, García-Pérez ML, Bilotta F. Intraoperative monitoring of cerebral oximetry and depth of anaesthesia during neuroanesthesia procedures. Curr Opin Anaesthesiol. 2016;29(5):576–81.
17. Wik L. Near-infrared spectroscopy during cardiopulmonary resuscitation and after restoration of spontaneous circulation: a valid technology? Curr Opin Crit Care. 2016 Jun;22(3):191–8.
18. Parnia S, Yang J, Nguyen R, Ahn A, Zhu J, Inigo-Santiago L, et al. Cerebral oximetry during cardiac arrest: a multicenter study of neurologic outcomes and survival. Crit Care Med. 2016;44(9):1663–74.
19. Genbrugge C, De Deyne C, Eertmans W, Anseeuw K, Voet D, Mertens I, Sabbe M, Stroobants J, Bruckers L, Mesotten D, Jans F, Boer W, Dens J. Cerebral saturation in cardiac arrest patients measured with near-infrared technology during pre-hospital advanced life support. Results from Copernicus I cohort study. Resuscitation. 2018;129:107–13.
20. Goldman S, Sutter F, Ferdinand F, Trace C. Optimizing intraoperative cerebral oxygen delivery using noninvasive cerebral oximetry decreases the incidence of stroke for cardiac surgical patients. Heart Surg Forum. 2004;7:E376–81.
21. Slater JP, Guarino T, Stack J, Vinod K, Bustami RT, Brown JM 3rd, et al. Cerebral oxygen desaturation predicts cognitive decline and longer hospital stay after cardiac surgery. Ann Thorac Surg. 2009;87:36–44.
22. Murkin JM, Adams SJ, Novick RJ, Quantz M, Bainbridge D, Iglesias I, et al. Monitoring brain oxygen saturation during coronary bypass surgery: a randomized, prospective study. Anesth Analg. 2007;104:51–8.
23. Heringlake M, Garbers C, Käbler JH, Anderson I, Heinze H, Schön J, et al. Preoperative cerebral oxygen saturation and clinical outcomes in cardiac surgery. Anesthesiology. 2011;114:58–69.

24. Bickler P, Feiner J, Rollins M, Meng L. Tissue oximetry and clinical outcomes. Anesth Analg. 2017;124(1):72–82.
25. Gregory A, Kohl BA. Con: near-infrared spectroscopy has not proven its clinical utility as a standard monitor in cardiac surgery. J Cardiothorac Vasc Anesth. 2013;27:390–4.
26. Pisano A. Can we claim accuracy from a regional near-infrared spectroscopy oximeter? Anesth Analg. 2016;122(3):920.
27. Bickler PE, Feiner JR, Rollins MD. Factors affecting the performance of 5 cerebral oximeters during hypoxia in healthy volunteers. Anesth Analg. 2013;117:813–23.
28. Tomlin KL, Neitenbach AM, Borg U. Detection of critical cerebral desaturation thresholds by three regional oximeters during hypoxia: a pilot study in healthy volunteers. BMC Anesthesiol. 2017 Jan 13;17(1):6.
29. Pellicer A, Bravo MC. Near-infrared spectroscopy: a methodology-focused review. Semin Fetal Neonatal Med. 2011;16(1):42–9.
30. Watzman HM, Kurth CD, Montenegro LM, Rome J, Steven JM, Nicolson SC. Arterial and venous contributions to near-infrared cerebral oximetry. Anesthesiology. 2000;93(4):947–53.
31. Statkiewicz Sherer MA, Visconti PJ, Russell Ritenour E, Haynes K. Radiation protection in medical radiography. 7th ed. The Netherlands: Elsevier Mosby; 2014. p. 310.
32. Aston D, Rivers A, Dharmadasa A. Equipment in anaesthesia and critical care. Branbury: Scion Publishing; 2014. p. 178.
33. Langton JA, Hutton A. Respiratory gas analysis. Contin Educ Anaesth Crit Care Pain. 2009;9(1):19–23.
34. Szocik JF, Barker SJ, Tremper KK. Fundamental principles of monitoring instrumentation. In: Miller RD, editor. Miller's anesthesia. 6th ed. Philadelphia, PA: Elsevier Churchill Livingstone; 2005. p. 1191–225.
35. Walker J. The flying circus of physics. 2nd ed. New York: Wiley; 2007. p. 293.
36. Ash C, Dubec M, Donne K, Bashford T. Effect of wavelength and beam width on penetration in light-tissue interaction using computational methods. Lasers Med Sci. 2017;32(8):1909–18.

Sunsets and Optical Fibers: A Bit of Geometrical Optics

18

Contents

Where's Physics

Daily life	How we see the objects around us Lenses, Glasses, Binoculars, Magnifiers, etc. Straws (apparently) bent in our drinks Imaginary puddles and other illusions Rainbows The setting sun has already set! Fogged up glasses or diving masks Fiber optic Internet connection
Physics involved	The "ray model" of light Reflection, refraction, and dispersion of light Fermat's principle (of least time) Snell's law (of refraction) Total internal reflection Surface tension (see Chap. 6)
Clinical practice	Fiber optic devices (bronchoscopes, laryngoscopes) Why endoscopic devices or personal protection equipment get fogged up (and how to reduce or prevent this problem)

A. Pisano, *Physics for Anesthesiologists and Intensivists*,
https://doi.org/10.1007/978-3-030-72047-6_18

18.1 Light as a Set of "Rays": Reflection and Refraction

In Chap. 16, we discussed the wave nature of light. However, we also mentioned that thinking at light as a set of "independent" rays traveling in straight lines (according to the so-called *ray model of light*) is a good approximation in several circumstances. The part of physics which deals with light in this way is called *geometrical optics* (and is the oldest part of optics, that is the study of light). In this chapter, we will see a few concepts and laws of geometrical optics which will allow us to understand how *optical fibers* like those contained in a bronchoscope work. We will also take the opportunity to discuss the physics behind the fogging of a bronchoscope or of personal protective equipment (PPE) such as goggles.

Let's start from a fundamental law of optics: the *Fermat's principle* (or *principle of least time*). A simple (though not entirely correct) formulation of Fermat's principle is the following: among all the possible paths which it might take to go from a point A to a point B, light *chooses* that which requires the *shortest time*. One might think that the path which requires the shortest time is *always* the shortest path (in terms of *space*), that is a straight line going from A to B, but it is not so. The reason is that the speed of light *varies* according to the medium in which it travels (see Table 18.1); accordingly, if the points A and B are located within two different media in which light speed is different, light makes appropriate deviations so as to *increase* the distance traveled in the medium where it travels faster and *decrease* the distance traveled in the one where it travels slower in order to comply with Fermat's principle. A particularly suggestive analogy (provided by Richard Feynman [1], one of the greatest physicists ever) is that of a lifeguard who has to save someone who is drowning in a body of water that is not in front of his/her current position: if (s)he can run faster than (s)he can swim, (s)he should run a certain stretch of the beach before diving, instead of immediately diving, in order to reach the victim as soon as possible (everything will be clearer below).

As mentioned in Chaps. 14 and 16, the *speed of light* in vacuum (symbolized c) is about 300,000 km/s (or 186,000 mi/s). However, light speed v through any (transparent) material medium is lower. The ratio of the speed of light in vacuum to the speed of light in any medium is called the *index of refraction* n of that medium (we will see shortly why it is called that):

Table 18.1 Speed of light (approximated to the nearest thousand) and index of refraction for some transparent media

Medium	Speed of light, km/s (mi/s)	Index of refraction (n)
Vacuum	300,000 (186,000)	1 (exactly)
Air	300,000 (186,000)	1.0003 (approximated to 1)
Water	225,000 (140,000)	1.33
Glass	200,000 (124,000)	1.5
Polystyrene plastic	189,000 (117.000)	1.59
Diamond	124,000 (77,000)	2.42

$$n = \frac{c}{v} \tag{18.1}$$

Table 18.1 reports the speed of light and the index of refraction for some materials. Usually, when light passes from a medium to another *with a different index of refraction* (i.e., in which its speed changes), part of the incident ray is *reflected* from the boundary between the two media, while part of the light is *refracted*, i.e., it continues through the second medium but a change in its direction occurs at the boundary (see Fig. 18.1). Both the reflected and the refracted ray lie in the same plane of the incident ray.

Now let's see how the direction of the reflected ray and that of the refracted ray are related to the direction of the incident ray.

18.1.1 The Law of Reflection

The relationship between the direction of the incident ray and that of the reflected ray (*law of reflection*) is very simple; the angle of reflection (i.e., the angle that the reflected ray forms with the normal to a plane tangent to the boundary) is always equal to the angle of incidence (i.e., the angle that the incident ray forms with the same normal). Hence, in Fig. 18.1, it is:

$$\theta_1 = \theta_1' \tag{18.2}$$

Equation (18.2) is a direct consequence of Fermat's principle; in other words, the law of reflection is nothing but the principle of least time applied to reflection. The simple geometrical reasoning used by Feynman in its landmark "lectures on physics" [1] allows us to realize this without resorting to boring equations and derivatives. Figure 18.2 shows a light ray traveling from a point A to a point B after hitting a mirror at point C (so that the angle of incidence is equal to that of reflection according to the law we have just mentioned). It is not difficult to prove that if light

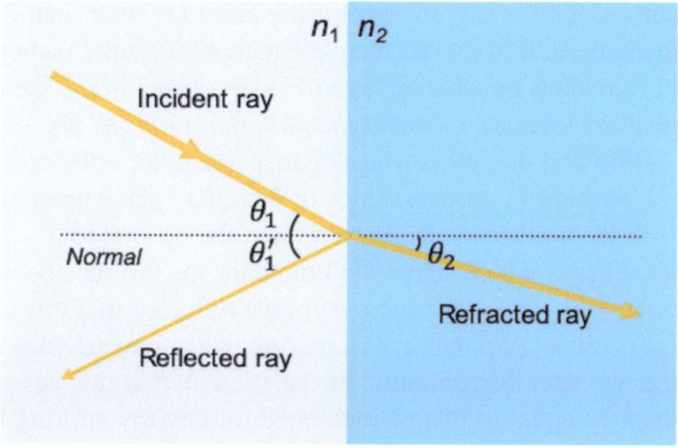

Fig. 18.1 Reflection and refraction of a light ray when it passes from a medium with index of refraction n_1 to a medium with index of refraction $n_2 > n_1$ (i.e., from a medium in which it travels faster to a medium in which it travels slower)

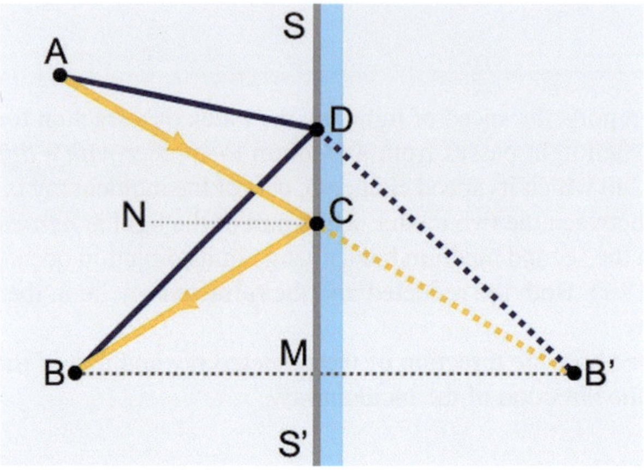

Fig. 18.2 Equivalence between the law of reflection and Fermat's principle. Line BB' meets the mirror plane at point M along the line SS'. Note that, since BM = MB' and B $\hat{M}$ D is a right angle, it is CB=CB' and DB=DB'. Moreover, the angle B $\hat{C}$ M is equal to the angle B′ $\hat{C}$ M and, hence, to the angle A $\hat{C}$ S; accordingly, the angles A $\hat{C}$ N and B $\hat{C}$ N that AC and CB form with the normal to SS' are also equal to each other since they are the complementary angles (see Box 1.1) of two equal angles, i.e., A $\hat{C}$ S and B $\hat{C}$ M, respectively. According to the triangle inequality theorem (see text or, again, Box 1.1), the path ACB' (and, hence, ACB) will be shorter than the path ADB' (and, hence, ADB) for any location of the point D (different from C) along the line SS'. Since the speed of light is constant, the path ACB (which, as shown geometrically, corresponds to the path for which the angle of incidence and the angle of reflection have the same width) is also that which requires the least possible time

would hit the mirror anywhere else (e.g., at point D), it would take more time to get from A to B. Let B′ be a point located behind the mirror at the same distance of B from it along the normal to the mirror surface; evidently, the distance CB is equal to the distance CB', and the fastest way to get from A to B′, *if speed is constant*, is a straight line that passes through the point C. In fact, since the sum of the lengths of any two sides of a triangle is always greater than the length of the third side (according to the so-called *triangle inequality theorem*), the path ACB' is shorter than the path ADB' for any position of the point D (other than C) along the line SS'. Hence, the path ACB is shorter than any path ADB with D other than C and, since the speed of light does not change (as reflection occurs in the same medium in which the incident ray travels), *it is also the path that requires the least time*.

Flat and smooth surfaces cause *specular reflection*: parallel incident rays give rise to parallel reflected rays. In fact, the "plane tangent to the boundary" coincides with the surface itself. Accordingly, the normal to this plane has the same direction for all rays which strike the boundary (as the surface is flat and without irregularities). This is how we see our image reflected in a mirror. Conversely, irregular surfaces reflect parallel rays in random directions (*diffuse reflection*): in fact, although the law of reflection must be satisfied also in this case (i.e., the angle of reflection must be equal to that of incidence for any ray striking the surface), the normal has different directions at different points due to the irregularities of the surface, so that parallel rays will form many different angles with the normal, giving rise to reflected rays in many different directions. We see most of the objects that surround us thanks to the diffuse reflection of light from them.

18.1.2 The Law of Refraction (Snell's Law)

As mentioned, light modifies its direction (i.e., it is refracted) when passing from a medium to another so as to "save time." Accordingly, it can be easily guessed that the extent of this deviation depends on how much the speed of light changes between the two media. In particular, the angle of refraction θ_2, that is the angle that the refracted ray forms with the normal to a plane tangent to the boundary between the two media (see Fig. 18.1), is related to the angle of incidence θ_1 by the following equation:

$$n_1 \sin\theta_1 = n_2 \sin\theta_2 \tag{18.3}$$

where n_1 and n_2 are the indexes of refraction of the two media. The refracted ray bends towards the normal (as in Fig. 18.1) if $n_2 > n_1$ and away from the normal if $n_2 < n_1$. Of course, the ray keeps going straight if $n_2 = n_1$ (in this case, in fact, it is sin θ_1 = sin θ_2 and, accordingly, $\theta_1 = \theta_2$); in other words, light is not refracted when it passes from a medium to another with the same index of refraction.

Equation (18.3) is the so-called *Snell's law of refraction*, also referred to as *Snell-Descartes law*, *Descartes law*, or simply *law of refraction*. It can also be written like this:

$$\frac{\sin\theta_1}{\sin\theta_2} = \frac{n_2}{n_1} \tag{18.4}$$

or, remembering the definition of index of refraction (Eq. 18.1):

$$\frac{\sin\theta_1}{\sin\theta_2} = \frac{n_2}{n_1} = \frac{c / v_2}{c / v_1} = \frac{c\,v_1}{v_2 c} = \frac{v_1}{v_2} \tag{18.5}$$

That is, the ratio between the angle of incidence and that of refraction is equal to the ratio of the speed of light v_1 in the first medium to the speed of light v_2 in the second medium. As with the law of reflection, the equivalence between Snell's law and Fermat's principle can be demonstrated geometrically (but we're not going to do it here because it's just a little bit more complicated) [1–3]. Hence, the ratio between the angles of incidence and refraction is simply the one which allows light to travel from point to point in the shortest possible time according to its different speed in different media. Similarly, in order to reach the victim as soon as possible, the aforementioned lifeguard should choose the angle at which to approach the shore from its position using Snell's law. However, (s)he should exactly know his/her own speed on the sand and in the water and have paper, pen, and a protractor with him/her; it is probably better that (s)he starts running without thinking too much!

Snell's law can also be derived describing light as a wave, but this would require the introduction of further notions and "principles" which is not the case to address here; just know that the ability of light to "choose" the most convenient path does not imply that it has a conscience; there is its wave nature (and quantum physics) behind all this!

Many everyday phenomena are due to refraction of light [1, 4]. A straw in a drink looks bent because the light that reaches our eyes from the part immersed in the drink is refracted when it passes from liquid to air, while that coming from the part that remains out is not. As shown in Fig. 18.3, when we admire the setting sun, the sun has already set! In fact, its light reaches us after being refracted when entering the atmosphere, but to an observer it seems to come from the backward extension of the refracted ray. The mirage of a lake in a desert or of a puddle ahead of us along a roadway (the so-called oasis mirage), which disappears when we get to where it should be, is caused by the refraction of light coming from the sky due to the gradual decrease of air temperature with altitude; in fact, (1) air is warmer the closer it is to the ground and (2) warmer (and, accordingly, less dense) air has a lower index of refraction. Hence, as they "descend" to the ground, the light rays are more and more refracted upwards (remember that the refracted ray is bent away from the normal when entering a medium with *lower* index of refraction), until they are intercepted by the eyes of someone driving along the road, to whom they will seem to come from the ground (that is, as in the example of Fig. 18.3, from the backward extension of the rays which reach his/her eyes). The apparent position of an object under the sea, which seems less deep of the actual position, is another consequence of light refraction. Finally, (but the list could go on and on [4]), a rainbow is produced by the combination of three optical phenomena: refraction, reflection, and (*chromatic*) *dispersion* (i.e., the "separation of colors" due to the wavelength-dependency of the index of refraction) by the raindrops.

A lot of objects of common use also rely on light refraction. A curved (usually *spherical*) shape of the boundary between different transparent media is exploited to build converging (convex) lenses, in which light rays entering parallel to the axis are all refracted towards a single point (called *focal point* or *focus*) at the other side, or diverging (concave) lenses, in which rays entering parallel to the axis diverge

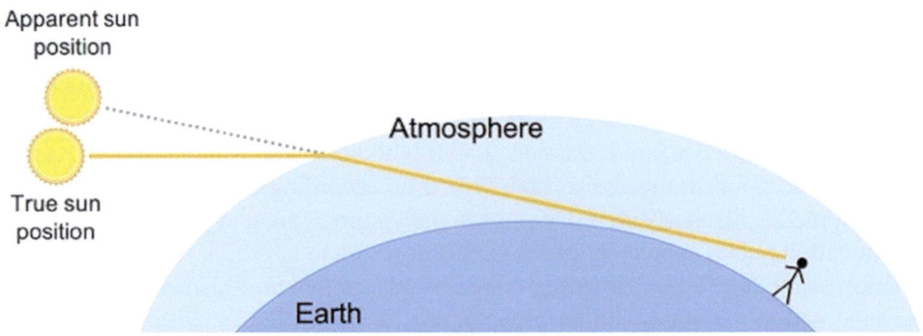

Fig. 18.3 True and apparent position of the sun near the horizon (see text)

when exiting from the other side as if they came from a single point (the so-called *virtual focus*). For example, a converging lens is a *magnifier* when used at an appropriate distance, while a combination of converging and diverging lenses is often used in instruments such as cameras or eyepieces for microscopes or (Galilean) telescopes. The anti-glare system of rearview mirrors in our cars is another example of a daily object which uses light refraction [4].

Now we are ready to see how an optical fiber works.

18.2 Total Internal Reflection and Optical Fibers

As stated above, the passage of a light ray from a transparent medium to another (also transparent) with a different index of refraction gives rise, *usually*, to a reflected ray and a refracted ray. Since "usually" does not mean "always," let's see when this does not happen. Figure 18.4 shows some light rays traveling through a medium with index of refraction n_1 and heading towards a medium with index of refraction n_2 *lower* than n_1. Figure 18.4a is similar to Fig. 18.1; at the boundary between the two media, we have a reflected ray (with an angle of reflection identical to the angle of incidence) and a refracted ray. The only difference is that, since it is $n_2 < n_1$ (contrary to what we assumed in Fig. 18.1), the refracted ray bends away from the normal instead of towards it. Remembering a bit of trigonometry (see also Chap. 1), Eq. (18.4) (Snell's law) tells us that the angle of refraction θ_2 *increases with the angle of incidence* θ_1; in fact, for a constant ratio between n_2 and n_1, $\sin\theta_2$ must increase as

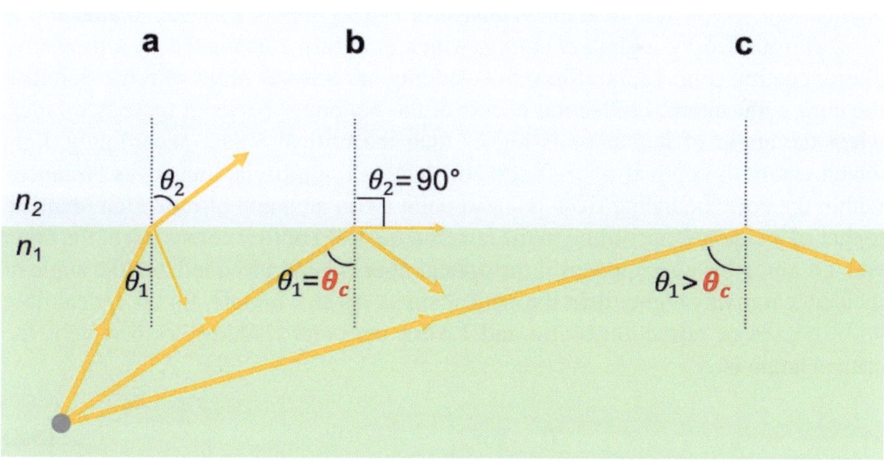

Fig. 18.4 Reflection and refraction of light rays passing from a medium with index of refraction n_1 to a medium with index of refraction $n_2 < n_1$ with different angles of incidence: (**a**) an angle of incidence θ_1 lower than a critical angle θ_c gives rise to a reflected ray and a refracted ray (which bends away from the normal to the boundary); (**b**) with a critical angle of incidence θ_c, the refracted ray points along the boundary (not entering the medium with index of refraction n_2); (**c**) for an angle of incidence $\theta_1 > \theta_c$ there is no refracted ray (total internal reflection)

$\sin\theta_1$ increases and since the sine of an angle assumes increasing values between 0 and 1 for (increasing) widths comprised between 0 and 90°, then also θ_2 must increase with θ_1.

It is evident that the maximum angle of refraction is 90°, that is a refracted ray pointing along the boundary between the two media (with a greater angle, it would be a reflected ray!). The angle of incidence which gives an angle of refraction of 90° is called the *critical angle* θ_c (Fig. 18.4b). To find the value of the critical angle for the boundary between two media with indexes of refraction n_1 and n_2, it is sufficient to assume $\theta_1 = \theta_c$ and $\theta_2 = 90°$ in Eq. (18.3) (or 18.4):

$$n_1 \sin\theta_c = n_2 \sin 90^\circ \tag{18.6}$$

Remembering that sin 90° = 1 and resolving for θ_c, Eq. (18.6) becomes:

$$\theta_c = \sin^{-1}\left(\frac{n_2}{n_1}\right) \tag{18.7}$$

where $\sin^{-1}$ stands for arcsine, i.e., the inverse of sine (see Chap. 1, Sect. 1.3.1).

When the angle of incidence is greater than the critical angle θ_c, there is no refracted ray; this situation is called *total internal reflection* (Fig. 18.4c). Total internal reflection can only occur when the incident ray lies in the medium with *higher* index of refraction; in fact, since the maximum value for the sine of an angle is 1, Eq. (18.7) has no solutions for $n_2 > n_1$.

Total internal reflection is what allows *optical fibers* to carry light (and information in the form of electromagnetic radiation) over great distances with very small attenuation. A typical optical fiber consists of a glass fiber of a few μm diameter (the *core*) surrounded by a glass *cladding*, which is in turn surrounded by a protective plastic coating (Fig. 18.5a). Since the cladding has a lower index of refraction than the core, total internal reflection occurs at the boundary between these two layers when the angle of incidence is higher than the critical angle; accordingly, light which enters the optical fiber at one end (with an appropriate angle) is "trapped" within the core, bouncing from point to point (with an angle of reflection identical to that of incidence, according to the law of reflection) until it comes out at the other end. Of note, this also happens if the optical fiber is bent, provided that the angle of incidence remains higher than the critical angle. As an example, for an optical fiber with indexes of refraction of 1.6 and 1.5 for core and cladding, respectively, the critical angle is:

$$\theta_c = \sin^{-1}\left(\frac{1.5}{1.6}\right) \approx 70^\circ \tag{18.8}$$

An infrared laser beam can travel for hundreds of kilometers, with negligible losses, along the optical fibers contained in the cables for fiber optic internet connection. Infrared radiation is used to encode the internet signal since the glass core of these fibers has the minimum absorption (and, hence, the minimum signal loss) in the infrared range of the electromagnetic spectrum (see Chap. 16).

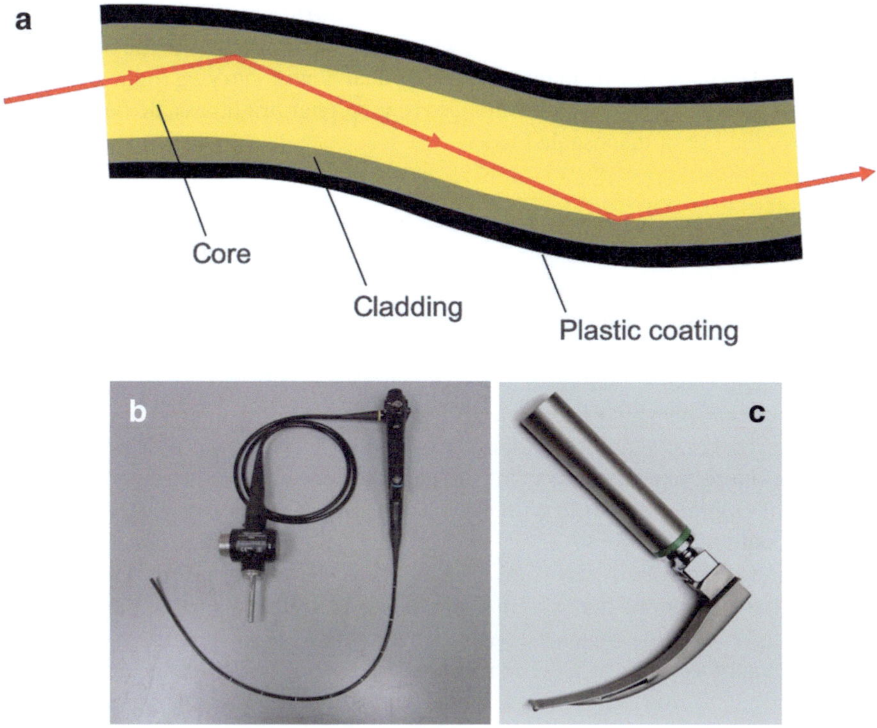

Fig. 18.5 Optical fibers: (**a**) schematic representation of an optical fiber showing total internal reflection within it; (**b**) a flexible bronchoscope containing thousands of optical fibers; (**c**) a fiber optic Macintosh laryngoscope

As we all know, several instruments commonly used in the intensive care unit or in the operating room use optical fibers (see Fig. 18.5b, c). In particular, a *flexible bronchoscope* (Fig. 18.5b) consists of two thin bundles of optical fibers: one carries light from a light source to the areas to be explored (so as to illuminate them without the need for a light bulb at the tip of the device), while the other one carries the reflected light from these areas up to the eyepiece lens, so as to view the images. The light rays reflected (diffusely) from the structures to be visualized enter the optical fibers bundle after being *focused* (i.e., all refracted towards the same point) by means of an *objective lens,* positioned at the tip of the bronchoscope, which allows to maximize (with a good compromise) both light collection and brightness. In other words, the objective lens collects all the light that is let into the optical fibers (by means of refraction) at an angle for which total internal reflection occurs; in this way, no light rays are lost (otherwise, a ray entering the core of the optical fiber so as to "hit" the core-cladding boundary with an angle of incidence lower than the critical angle would be partially refracted through the cladding and it would not arrive "intact" at the other end of the optical fiber). This is all wonderful as long as the bronchoscope doesn't fog up! (See Box 18.1).

Finally, Fig. 18.5c shows a *fiber optic (Macintosh) laryngoscope*, in which the light bulb is housed in the handle instead of on the blade (such as in the older "conventional" laryngoscopes), thanks to the optical fibers carrying light along the blade: this allows, among other advantages (e.g., higher brightness, no heat transfer to tissues), the use of disposable blades.

Box 18.1: Diving Masks, Goggles, and Bronchoscopes that Fog Up: Surface Tension and Geometrical Optics

Divers are familiar with the problem of mask fogging, and often they also know how to avoid it, although in a not too elegant way: that is, by spitting into the mask and spreading saliva everywhere inside it! [5]. In the very particular period in which I'm writing, that is in the midst of the coronavirus disease 2019 (COVID-19) pandemic [6, 7], this has become a common issue also for intensivists (and physicians in general) who are forced to work wearing tight *goggles*, similar to a diving mask, together with other personal protective equipment. Moreover, as anyone who has ever used a bronchoscope knows, also these instruments may fog up, thus preventing vision. The physics we have discussed here and in Chap. 6 provides an explanation (and some solutions) for all this. When we wear a diving mask or protective goggles, our sweat evaporates and, as a consequence, water vapor accumulates within the mask/goggles. However, since the outer part of the device is in contact with water (if we are divers) or air (if we are physicians wearing goggles) which is colder than our skin, water vapor condenses (i.e., it becomes liquid) on the plastic surface. Due to *surface tension*, water forms a lot of (spherical) tiny drops (see Sect. 6.1.1), which make light coming from surrounding objects being *refracted*, *reflected*, and (sometimes) *dispersed* in random directions, thus preventing us from seeing clearly [5]. The same occurs when the water vapor contained in the (humidified) air in the airways meets the colder surface of the objective lens on the tip of a bronchoscope; if the bronchoscope were at the body temperature, it would not fog up. This is why, once, otolaryngologists used to heat up the laryngeal mirror before performing indirect laryngoscopy (and, similarly, pre-heaters for bronchoscopes exist). As for diving masks, goggles, and bronchoscopes, however, what can be usually done to prevent (or at least reduce) fogging, is to *reduce surface tension*, so that water forms an even layer instead of droplets which will behave as many lenses [5, 8]. Although specific anti-fogging solutions containing *surfactants* (which reduce surface tension) are available both for bronchoscopes [8, 9] and for glasses or diving masks, you can make do with what you find: ethyl alcohol may be suitable for a bronchoscope, a small amount of hand soap can be spread inside the goggles [10], while a little saliva may work for a diving mask!

References

1. Feynman RP, Leighton RB, Sands M. The Feynman lectures on physics: the new millennium edition. New York: Basic Books; 2011. p. 26.1–8.
2. Golomb M. Elementary proofs for the equivalence of Fermat's principle and Snell's law. Am Math Mon. 1964;71(5):541–3.
3. Helfgott H, Helfgott M. A noncalculus proof that Fermat's principle of least time implies the law of refraction. Am J Phys. 2002;70(12):1224–5.
4. Walker J. The flying circus of physics. 2nd ed. New York: Wiley; 2007. p. 242–304.
5. Czerski H. The storm in a teacup. New York: W. W. Norton & Company; 2018. p. 61–85.
6. Pisano A, Landoni G, Verniero L, Zangrillo A. Cardiothoracic surgery at the time of the coronavirus disease-2019 pandemic: lessons from the east (and from a previous epidemic) for Western Battlefields. J Cardiothorac Vasc Anesth. 2020;34(9):2338–40.
7. Pisano A, Yavorovskiy A, Verniero L, Landoni G. Indications for tracheal intubation in patients with coronavirus disease 2019 (COVID-19). J Cardiothorac Vasc Anesth. 2021;35(5):1276–80.
8. Leonard RL, Terekhov AY, Thompson C, Erck RA, Johnson JA. Antifog coating for bronchoscope lens. Surf Eng. 2002;28(6):468–72.
9. Knauth A, Weiss M, Dave M, Frotzler A, Haas T. Comparison of antifog methods in endoscopy. What really helps. Anaesthesist. 2012;61(12):1036–44. Article in German
10. Malik SS, Malik SS. A simple method to prevent spectacle lenses misting up on wearing a face mask. Ann R Coll Surg Engl. 2011 Mar;93(2):168.

Part VIII

Sound Waves, Resonance, Ultrasonography

Oscillations and Resonance: Origin and Propagation of Sound, Children on the Swing, and Invasive Pressure Monitoring

19

Contents

Where's Physics

Daily life	Flying balloons and funny voice Musical instruments Children on the swing Trees broken by a weak wind Collapsing buildings
Physics involved	Origin and propagation of sound Fourier theorem (a few hints) Simple harmonic motion Hooke's law (spring force) Natural frequency Resonance
Clinical practice	A monster in the operating room (xenon anesthesia) Invasive arterial pressure monitoring: • Underdamping (resonance) • Overdamping • Natural frequency and damping coefficient • Signal processing

A. Pisano, *Physics for Anesthesiologists and Intensivists*, https://doi.org/10.1007/978-3-030-72047-6_19

19.1 Origin and Propagation of Sound

As mentioned in Chap. 16, sound is a mechanical wave, i.e., a periodic disturbance (or *oscillation*) which needs a material medium such as a gas, a liquid, or a solid in order to propagate (unlike electromagnetic waves, which can travel through a vacuum). In fact, what oscillates is the *density* (and, hence, the *pressure*) within the medium; accordingly, there cannot be a *sound wave* if there are not "particles" (e.g., air molecules) which can periodically come closer and move away with one another, i.e., undergo periodical *compression* and *rarefaction* (Fig. 19.1). Moreover, the medium must be *elastic*, i.e., its molecules must be able to change their reciprocal positions (which means a variation in density) when subjected to a "perturbation," but there must be some force that makes them return to the initial positions once the perturbation is removed. Another difference with respect to electromagnetic waves is that, as shown in Fig. 19.1, sound waves are *longitudinal waves*, i.e., they travel parallel to the direction in which the oscillation occurs. All this is usually summarized by saying that *a sound wave is a longitudinal wave* (*of compression and rarefaction,* or a "*pressure wave*") *which travels through an elastic medium.*

Sounds are produced by something, such as a guitar string, a drum-skin, or our vocal cords, that vibrates periodically (conversely, what we call "noise" is an acoustic disturbance which propagates through an elastic medium *without* a precise periodicity). Although periodic, sounds we hear commonly, including our voice, have not a *sinusoidal* (i.e., *sine* or *cosine*) waveform, but are generally more complex waves. However, according to the so-called *Fourier theorem*, any periodic wave can be described as the sum of sinusoidal waves whose frequencies (see Chap. 16) are integer multiples of its frequency. Accordingly, a sound can be always considered as the sum of a certain number of *pure tones*, called *harmonics*, each described by a sinusoidal wave. The wave with the lowest frequency is referred to as the

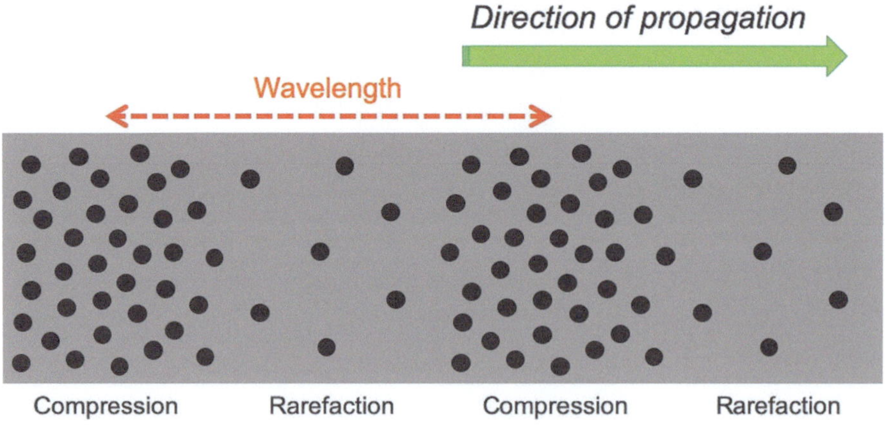

Fig. 19.1 The molecules of a material medium which is crossed by a sound wave periodically approach and move away from each other (compression and rarefaction) along the direction of wave propagation

fundamental mode or *first harmonic*, while the other sounds, whose frequencies are integer multiples of the first harmonic (see below), are called second harmonic, third harmonic, and so on (Fig. 19.2). This "set" of waves forms a *Fourier series* (or *harmonic series*). The study of Fourier series is known as *Fourier analysis*: as mentioned in Chap. 9, this mathematical tool is used for signal processing during invasive pressure monitoring and allows monitors to display an accurate representation of the arterial pressure waveform [1–4].

The main properties of waves (amplitude, frequency, wavelength, and speed of propagation) have been already discussed in Chap. 16.

Similarly to electromagnetic waves, the amplitude of a sound wave expresses the amount of *energy* carried by the wave, i.e., its *intensity*; a higher amplitude corresponds to a louder sound.

The frequency of the first harmonic (called *fundamental frequency*) indicates the "pitch" of the sound; higher frequencies are heard as more acute sounds. We can hear sounds with a frequency between 20 and 20,000 Hertz (1 Hz = 1 "oscillation per second"), while "sound" waves with higher frequency (> 20 kHz) are called *ultrasounds*, whose clinical use is known to each of us (see Chap. 20).

As discussed in Chap. 16, frequency f and wavelength λ are correlated with each other according to the following equation (which is analogous to Eq. 16.3):

$$f = \frac{v}{\lambda} \tag{19.1}$$

where v is the wave speed, namely, in this case, the *speed of sound*, which depends on both the density and the elastic properties of the medium in which sound propagates (see below).

Finally, the *timbre* (or "quality") of a sound, i.e., what distinguishes my voice from that of someone else, or a musical note played on a clarinet from the same note

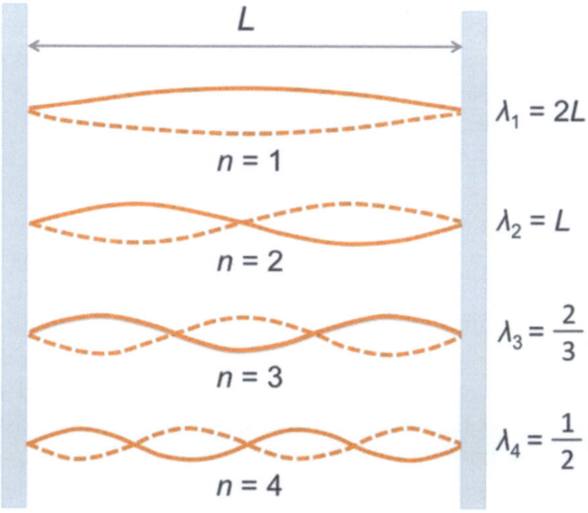

Fig. 19.2 The first four harmonics (or modes) on a string of length L (see text). The wavelength λ of each nth harmonic is 2 L/n. Evidently, if f is the frequency of the first harmonic, the frequency of each nth harmonic is nf (i.e., $2f$ for the second harmonic, 3f for the third harmonic, and so on). n harmonic number

played on a piano, depends on the percentage of the different harmonics which are present, i.e., on the sound *waveform*.

The next section is between serious and humorous; we will see something more about the origin and propagation of sound in order to explain something "strange" which might (theoretically) occur during a xenon anesthesia. However, it will also give us the opportunity to introduce the notion of *resonance*, which is addressed in Sect. 19.2 together with other important topics such as (simple) harmonic motion and natural frequency; all this will help us to deeply understand what leads to underdamping of invasive pressure monitoring systems.

We will continue to talk about wave physics in Chap. 20, which is mainly dedicated to *Doppler Effect* and its application to both ultrasonography (e.g., echocardiography) and noninvasive cardiac output monitoring.

19.1.1 A Monster in the Operating Room

If you should happen to induce *xenon anesthesia* [5–8] through a face mask (something quite unusual, however) and hear spooky sounds coming from the patient, you must not think the patient is turning into a monster!

Everyone knows that if you inhale the helium contained in a "flying balloon" (see Chap. 3), your voice becomes very similar, both in its rapidity and in its unusually high tone, to that of the Disney's character Donald Duck [9, 10]. Let's see why.

The speed of sound v through any medium depends on the *density* d of that medium, according to the following equation:

$$v = \sqrt{\frac{B}{d}} \tag{19.2}$$

where B is the so-called *bulk modulus*, which represents the elastic properties of the medium. In practice, B indicates how much the volume of a material varies as the pressure exerted on it changes; the lower the bulk modulus of a material, the higher its *compressibility*. Accordingly, the speed of sound is much lower in gases, which are highly compressible, than in (almost incompressible) liquids.

Since the density of helium is lower than that of air [7, 11], the speed of sound through helium (965 m/s) is much higher than through air (343 m/s at 20 °C); this explains why words seem to come out faster [10], but not why the voice looks higher.

As mentioned, our voice, as well as most sounds we hear in everyday life, is not formed by a single (sinusoidal) sound wave, but by a "mixture" of harmonics. Consider the example of a guitar string fixed at its two ends, as shown in Fig. 19.2. Imagine a sinusoidal wave traveling along the string from the left to the right. Once the wave has reached the right end, it undergoes *reflection* and starts to travel in the opposite direction, until it reflects again at the left end. In a short time, there will be many waves that *interfere* with one another. While most of these waves cancel each other out, only those with certain frequencies add up to form a *standing wave* (or *stationary wave*), in which the points of minimum and maximum amplitude are fixed in the space. These frequencies are called *resonant frequencies*, and the standing wave is said to form at *resonance* (something that must be avoided during invasive pressure monitoring, as discussed in Sect. 19.2).

In order to produce a standing wave on a string of length L, a wave must have one of the following wavelengths:

$$\lambda = \frac{2L}{n} \tag{19.3}$$

where $n = 1, 2, 3,\ldots$ is the so-called *harmonic number*, i.e., $n = 1$ for the first harmonic (fundamental mode), $n = 2$ for the second harmonic, $n = 3$ for the third harmonic, and so on. In other words, Eq. (19.3) provides the wavelengths of the different waves (harmonics) that form an harmonic series; the wavelength of the first harmonic ($n = 1$) will be equal to twice the length of the string, that of the second harmonic ($n = 2$) will be equal to the length of the string, that of the third harmonic ($n = 3$) will be equal to two-third of the length of the string, and so on (see Fig. 19.2).

Equation (19.3) also applies to a pipe of length L (with two open ends) filled with air; in this case, a standing wave is produced by sound waves of appropriate wavelengths (i.e., equal to 2 L, L, 2/3 L, etc.) which propagate through the air within the pipe. The corresponding frequencies f at which the pipe *resonates* (resonant frequencies) can be found according to Eq. (19.1) (after substituting for λ from Eq. 19.3):

$$f = \frac{v}{\lambda} = n\frac{v}{2L} \tag{19.4}$$

where v is the speed of sound *through air*. Hence, the frequency of each harmonic n depends on the length L of the pipe and on the speed of sound through the gaseous medium contained in it.

Substituting for v from Eq. (19.2) (which provides the speed of sound through any medium), Eq. (19.4) becomes:

$$f = \frac{n}{2L}\sqrt{\frac{B}{d}} \tag{19.5}$$

where B and d are, respectively, the bulk modulus and the density of the gas that fills the pipe; hence, *the lower the density, the higher the frequency* of each harmonic (including the first harmonic, on which depends, as mentioned, the perceived "pitch" of the sound).

When our vocal cords vibrate to produce a sound, the oropharynx acts as a resonating cavity, as in a pipe organ, in a clarinet, and so on. Therefore, the vibration of vocal cords will produce a number n of harmonics, whose frequencies depend both on the size and shape of the oropharynx and on the speed of sound through the gas contained in it (which, in turn, depends on the density of that gas). When you inhale helium, the shape of your oropharynx does not change, but it is filled with helium instead of air; accordingly, the speed of sound v in it will be higher and, consequently, the harmonic frequency will be higher and the voice will look more acute [11].

Xenon has, conversely, a density about three times higher than that of air [7], so that the speed of sound through it is lower as compared with that through air; phonation while breathing xenon will therefore result in a slow and low-pitch voice, like that of a monster.

19.2 Children on the Swing and Invasive Pressure Monitoring: Oscillations, Natural Frequency, and Resonance

The phenomenon of *resonance*, which we have mentioned in the previous section, does not concern sound waves only, but mechanical waves in general. Any object may oscillate if opportunely solicited. Of course, there are objects that can be seen oscillating much more easily than others; think, for example, at a tree or a road sign in a windy day, a pendulum, a swing with a happy child, or the Einstein's head of the bobblehead figurine on my desk. When left to itself, any object that oscillates does so at its own frequency, which is called the *natural frequency* f_N of the object; this is the frequency with which the Einstein's head will oscillate if I touch it once, or the (much lower) frequency with which a child will oscillate on the swing after a single push. However, an object may also undergo a *driven oscillation* when a periodic external force is applied to it (as when you continue pushing a child on a swing). When the frequency of such periodic external solicitation, which is referred to as *driving frequency* f_D, matches the natural frequency f_N of the object, the object *resonates*, i.e., responds to such a solicitation by oscillating with large amplitude. In other words, an object undergoes resonance when it is:

$$f_D = f_N \tag{19.6}$$

that is, when it is prompted to oscillate by an external force applied with a frequency which is equal to the frequency at which the object tends to oscillate if left to itself (natural frequency). Everyone knows that if you want to make a child happy on a swing, you have to push him/her with a precise rhythm: if you push "too soon," the amplitude of the oscillations will shrink immediately. This is because the frequency of your pushes must match the natural frequency of the swing in order to make it resonate, that is, to make it oscillate widely. As another example, a tree may sway exaggeratedly (and, accordingly, be broken) due to a rather weak wind if the mean frequency of the wind gusts approaches its natural frequency [12].

While resonance is very useful for musical instruments, in which the sound produced by something that vibrates (e.g., a string or a reed) is amplified and enriched with harmonics by *resonators* (such as a pipe or a sound box), it can be a problem for rise buildings, airplanes … and invasive pressure monitoring systems [2, 4, 13, 14]. The natural frequency of the wings of an airplane must be very far from the frequency of the vibrations produced by air turbulence; otherwise, they could intensely oscillate and even (dramatically) break. Similarly, if one of the harmonics which form the arterial pressure waveform (see Chap. 9 and Sect. 19.1 in this chapter) has a frequency close to the natural frequency of the transducer-tubing-catheter system, the system may resonate, i.e., its own (driven) oscillation will significantly add to the oscillation of blood pressure, leading to overestimation of systolic pressure and to underestimation of diastolic pressure (the so-called *underdamping*, as we will see below). In order to avoid resonance, the natural frequency of an invasive pressure monitoring system must be at least 8-10 times higher than the measured heart rate (that is the frequency of the first harmonic). For example, if heart rate is 90 beats per minute (i.e., 1.5 beats per second or 1.5 Hz), the natural frequency of the system must be at least 12–15 Hz. Although commonly used pressure transducing systems have much higher natural frequencies, several factors may considerably decrease the natural frequency of the system, leading to resonance (and, hence, to underdamping).

In general, the natural frequency of an object depends on several factors including its *length* and *stiffness*. The greater the length of an object is, the lower its natural frequency. A swing with longer chains has a lower natural frequency; accordingly, it oscillates with lower frequency and one needs to push less frequently to make it resonate (and make children happy). In 1995, a violent earthquake struck Mexico City, collapsing many intermediate-height buildings, but not the taller and shorter ones; this was because the frequency of the seismic waves matched the natural frequency of intermediate-height buildings, causing them to oscillate widely (due to resonance), while shorter and taller buildings, with higher and lower natural frequencies, respectively, were mostly spared [15].

To find out what factors affect the natural frequency of an invasive pressure monitoring system, possibly leading to resonance (and, hence, to underdamping), we have to take a small step back and introduce the simplest form of oscillation: *simple harmonic motion.*

19.2.1 Simple Harmonic Motion

Any motion that repeats regularly over time is called *harmonic* (or *periodic*) *motion.* A periodic motion which can be described as a sinusoidal (i.e., sine or cosine) function of time is called *simple harmonic motion* (SHM), and any object which moves (oscillates) in this way is referred to as a *simple harmonic oscillator.*

Figure 19.3a shows a mass m attached to one end of a spring, whose other end is attached to a fixed support. The surface on which the mass lies is assumed to be frictionless. When you push or pull the mass, it starts to oscillate around the equilibrium (rest) position, with maximum displacement $x = \pm M$. Such a system is the prototype of a simple harmonic oscillator (a pendulum is another example); that is, the mass m undergoes SHM when pushed or pulled. Hence, its position x over time can be described by a sinusoidal wave (Fig. 19.3b shows the graph of a sine function

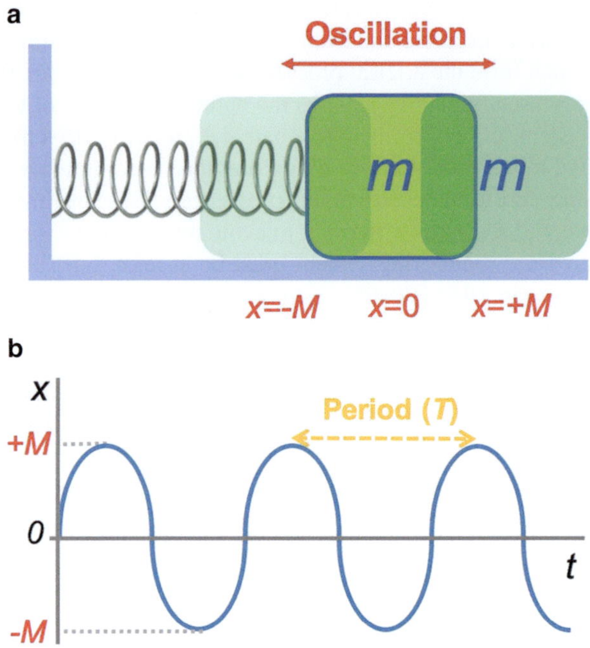

Fig. 19.3 A mass on a spring (**a**) is the prototype of a simple harmonic oscillator (see text). If pushed or pulled and then released, the mass *m* starts to oscillate with simple harmonic motion (SHM), that is with a motion which can be described as a sinusoidal function of time (**b**); in particular, the mass displacement *x* over time (*t*) is here graphically represented by a sine wave, but a cosine wave could be used as an alternative. In SHM, also speed and acceleration are sinusoidal (sine or cosine) functions of time (not shown). ±*M* Maximum displacement from the equilibrium position ($x = 0$)

like that of Figs. 1.5a and 16.2, but a cosine function such as that shown in Fig. 1.5b was just as correct); the mass continuously moves from the position +*M* to the position -*M* passing through the rest position ($x = 0$).

The study of the equations which describe SHM is of great importance since, as mentioned in Sect. 19.1, more complex oscillations such as sound waves or the arterial pressure waveform can be always "decomposed" into a series of sinusoidal waves (according to the *Fourier theorem*). Below we discuss just a few of these equations that interest us.

During SHM, *acceleration a* is always opposite to displacement *x*:

$$a = -\omega^2 x \tag{19.7}$$

In other words, the mass *m decelerates* as its displacement increases, and vice versa; that is, its *speed* continuously decreases as it moves from the equilibrium position ($x = 0$) to the points of maximum displacement ($x = +M$ and $x = -M$); in these two points, speed is zero (in fact, the mass must stop before it can go back); once its direction is reversed, the mass starts to *accelerate* up to a maximum speed at the equilibrium position. Acceleration and displacement are correlated to each other by the constant ω^2, that is the square of a quantity called *angular frequency* ω, which is defined as:

$$\omega = f_N \, 2\pi \tag{19.8}$$

where f_N is the natural frequency of the system.

We know from Newton's second law (see Chap. 14) that a body accelerates only when a *force* is acting on it. In this case, the force that accelerates the mass m is the so-called *spring force*, i.e., the force exerted by the spring when it is lengthened or shortened as compared with its *equilibrium length*. This force is a *restoring force*, that is the force that returns any *elastic* object to its equilibrium position. The value of the force exerted by the spring in Fig. 19.3 when the mass m is moved away from its equilibrium position by pulling or pushing it (that is, when the spring is lengthened or shortened, respectively, as compared with its equilibrium length) is provided by *Hooke's law*. This law states that *the elastic deformation of an object is always proportional to the force which has caused such a deformation*:

$$x = \frac{F}{k} \tag{19.9}$$

where k is a constant, called *spring constant*, which is a measure of the *stiffness* of the spring. In practice, the stronger the force you apply, the greater the displacement x from the equilibrium position (i.e., the spring lengthening or shortening with respect to the equilibrium length) you get; the higher the stiffness of the spring, the greater the force you will have to apply to get a certain displacement x. In particular, from Eq. (19.9), the magnitude of the force which causes the displacement x is $F = kx$. The spring will respond to this deformation with a restoring force (spring force) $\vec{F_S}$ which is equal (in magnitude) and opposite (in direction) to F:

$$\vec{F_S} = -\vec{F} \tag{19.10}$$

or

$$\vec{F_S} = -k\vec{x} \tag{19.11}$$

We can then reformulate Hooke's law as follows: the spring force $\vec{F_S}$ is directly proportional to the displacement x from the equilibrium position and always acts in the opposite direction to the displacement itself (hence the minus sign). In other words, the more we lengthen or shorten the spring, the stronger will be the force that tends to bring it *back* to its equilibrium length. The arrows on F_S and x remind us that these quantities are vectors (see the beginning of Chap. 13).

Let's see what happens by putting all these equations together. According to Newton's second law (Eq. 14.1), it is:

$$F_S = m\,a \tag{19.12}$$

After substituting for a from Eq. (19.7), we can write:

$$F_S = m\left(-\omega^2 x\right) = -m\,\omega^2 x \tag{19.13}$$

According to Eq. (19.11) (Hooke's law), F_s can be replaced with $-k\,x$, and Eq. (19.13) becomes:

$$-k\,x = -m\,\omega^2 x \tag{19.14}$$

or

$$k = m\,\omega^2 \tag{19.15}$$

Resolving for ω, we can rewrite Eq. (19.15) as:

$$\omega = \sqrt{\frac{k}{m}} \tag{19.16}$$

or, after substituting for ω from Eq. (19.8):

$$f_N\,2\pi = \sqrt{\frac{k}{m}} \tag{19.17}$$

that is,

$$f_N = \frac{1}{2\pi}\sqrt{\frac{k}{m}} \tag{19.18}$$

Equation (19.18) provides the natural frequency of the mass-spring system illustrated in Fig. 19.3 as a function of its *mass* and *stiffness* (we assume that the spring is massless and that the only elastic object is the spring; accordingly, both the mass m and the spring constant k can be referred to the whole system). In the next section, this equation will help us to find out the factors on which the natural frequency of an invasive pressure monitoring system depends.

19.2.2 Natural Frequency and Resonance of an Invasive Pressure Monitoring System: Possible Causes of Underdamping

Roughly (leaving out many details), an invasive pressure monitoring system can be regarded as a simple harmonic oscillator like the mass-spring system described above [1, 2, 16–18]. To make things even easier, we can limit our attention to the fluid-filled tube which connects the arterial cannula (or any other catheter used to measure pressure somewhere) to the pressure transducer, and consider it as an elastic system which oscillates longitudinally, exactly as the mass-spring system of Fig. 19.3, under the "drive" of the pressure which is measured.

We can then try to write Eq. (19.18) for this fluid-filled tube. (Notice: if they don't scare you a lot of mathematical passages, quite easy however, below I will take you step by step along a simplified derivation of the equation for the natural frequency of an invasive pressure monitoring system; otherwise, you can directly jump to Eq. (19.33)).

Let's start by defining the *stiffness* of such a system as its *volume elasticity* E, that is the ratio of the pressure change ΔP which is produced by a volume change ΔV to ΔV itself:

$$E = \frac{\Delta P}{\Delta V} \tag{19.19}$$

This is a very familiar notion for us intensivists or anesthesiologists; in fact, it is none other than what we call *elastance* if the "tube" is an artery (*arterial elastance*) [19], or the inverse of what we call *compliance* (i.e., $\Delta V/\Delta P$) when referring to the lung. It is also roughly the same as the "bulk modulus" mentioned above.

Remembering the (gross) definition of pressure as force F per unit area A (see Chap. 9, Sect. 9.1.1), Eq. (19.19) can be rewritten as:

$$E = \frac{F}{A\Delta V} \tag{19.20}$$

where F is evidently the force which periodically "deforms" (that is, as we will see shortly, makes lengthen) the tube, namely the "periodic external force" which is responsible for the driven oscillation of the invasive pressure monitoring system (in practice, the force exerted "by the arterial pulse" if we are monitoring arterial pressure).

Since the tube is cylindrical, its cross-sectional area (on which the force F acts perpendicularly) is $A=\pi r^2$, where r is the radius of the tube (see Chap. 1). Accordingly, Eq. (19.20) becomes:

$$E = \frac{F}{\pi r^2 \Delta V} \tag{19.21}$$

As we assumed that the system oscillates longitudinally (i.e., in the direction of its length), the volume change ΔV is nothing but the volume of a cylinder with base A and height ΔL, that is:

$$\Delta V = A\Delta L \tag{19.22}$$

where A ($=\pi r^2$) is, again, the cross-sectional area of the tube, and ΔL is the variation in the tube length, which is the equivalent of the displacement x of the mass on the spring. After substituting for ΔV from Eq. (19.22), Eq. (19.21) becomes:

$$E = \frac{F}{\pi r^2\, A\Delta L} = \frac{F}{\pi r^2 \left(\pi r^2\right)\Delta L} \tag{19.23}$$

Since we are considering our fluid-filled tube exactly the same way as the mass-spring system of Fig. 19.3, we can apply the *Hooke's law* discussed in Sect. 19.2.1 (Eq. 19.9) to find the force F which appears in Eq. (19.23). That is, if the magnitude of the force which causes the displacement x of the mass m from its equilibrium position is $k\,x$, then the magnitude of the force F which causes the elongation ΔL of the tube is:

$$F = k\Delta L \tag{19.24}$$

Due to its elastic properties, the system consisting of the fluid-filled tube will respond to the deformation ΔL with an equal and opposite restoring force $-k\Delta L$ (Eq. 19.11) which is analogous to the spring force discussed above and is responsible for the (driven) oscillation of the system.

Hence, after substituting for F from Eq. (19.24), Eq. (19.23) becomes:

$$E = \frac{k\Delta L}{\pi r^2\left(\pi r^2\right)\Delta L} \tag{19.25}$$

where the two "ΔL" cancel each other out, giving:

$$E = \frac{k}{\pi r^2\left(\pi r^2\right)} \tag{19.26}$$

or, resolving for k:

$$k = E\pi r^2\left(\pi r^2\right) \tag{19.27}$$

Here, it finally comes into play Eq. (19.18), in which, according to Eq. (19.27), we can replace k to the numerator with $E\,\pi r^2\,(\pi r^2)$, thus obtaining:

$$f_{\mathrm{N}} = \frac{1}{2\pi}\sqrt{\frac{E\pi r^2\left(\pi r^2\right)}{m}} \tag{19.28}$$

Now let's deal with the denominator. Remembering the definition of *density* which is provided in Chap. 9 (Eq. 9.1) and neglecting the tube wall, we can say that the mass m of our fluid-filled tube is:

$$m = dV \tag{19.29}$$

where d is the fluid density and V is the volume of the "cylinder of fluid" which fills the tube. The latter is, evidently (see above):

$$V = AL = \left(\pi r^2\right)L \tag{19.30}$$

with A (= πr^2) being the base of the cylinder (whose radius is r) and L being its length. Putting Eqs. 19.29 and 19.30 together, we can write:

$$m = d\left(\pi r^2\right)L \tag{19.31}$$

After substituting for m from Eq. (19.31), Eq. (19.28) becomes:

$$f_{\mathrm{N}} = \frac{1}{2\pi}\sqrt{\frac{E\pi r^2\left(\pi r^2\right)}{d\left(\pi r^2\right)L}} \tag{19.32}$$

or, noting that one of the two "πr^2" to the numerator cancels out with the equal term to the denominator:

$$f_{\mathrm{N}} = \frac{1}{2\pi}\sqrt{\frac{E\pi r^2}{d\ L}} \tag{19.33}$$

An equation just like this or very similar to it is the formula for the natural frequency of an invasive pressure monitoring (transducer-tubing-catheter) system reported in several papers or book chapters [13, 16–18, 20]. According to Eq. (19.33), *adding too long tubing* (i.e., increasing L), *using small diameter arterial cannulas*/connection tubes (i.e., reducing r), and the presence of (small) *air bubbles*, which reduce the stiffness (i.e., the volume elasticity E), may considerably decrease the natural frequency of the system. This may lead to resonance, especially if heart rate is high. In fact, as discussed above, according to the Fourier theorem the arterial pressure waveform (like any other periodic wave, regardless of its complexity) can be represented as the sum of a set of simple sine waves (or "harmonics"), which is called a *harmonic series*. The first element of the harmonic series, or "first harmonic," has the lowest frequency (fundamental frequency), which corresponds to the sound pitch in the case of a sound wave (see Sect. 19.1) and to *heart rate* (HR) in the case of the arterial pressure waveform. The subsequent harmonics have increasing frequencies which are all integer multiples of the fundamental frequency. That is, if the frequency of the first harmonic is f, that of the second harmonic is $2f$, that of the third harmonic is $3f$, and so on (just have a look at Fig. 19.2). Since at least 6 to 10 harmonics are usually required for a reliable reproduction of the arterial pressure waveform by means of Fourier analysis (see above) [2], the frequencies involved may be high enough to match the system's natural frequency, especially if the latter is reduced due to one or more of the factors which appear in Eq. (19.33) and heart rate is high. For example, if HR is 60 beats per minute (i.e., 1 beat per second or 1 Hz), the eighth harmonic will have a frequency of 8 Hz; if the monitoring system has a natural frequency of, say, 22 Hz, there is low risk of resonance even in the presence of factors which reduce the natural frequency of the transducer-tubing-catheter system to, for example, 18 Hz. However, with a HR of 120 min^{-1} (= 2 Hz) , the frequency of the eighth harmonic will be 16 Hz, and the likelihood that this frequency could match the natural frequency of the system if we add, for example, a tube extension (thus further reducing the system's natural frequency), becomes concrete. As mentioned, if that happens the system resonates, i.e., it oscillates "exaggeratedly," and this translates into an overestimation of systolic pressure and in an underestimation of diastolic pressure. Why this is referred to as "underdamping," and why the same factors which reduce the natural frequency of the system may also lead to the opposite phenomenon (overdamping) is briefly discussed in Box 19.1.

Box 19.1: Damped Oscillations: Natural Frequency, Damping Coefficient, and Overdamping of Invasive Pressure Monitoring Systems

So far in this chapter we have talked about objects which, when opportunely solicited and then left to themselves, *start* to oscillate with a frequency (called natural frequency) which depends on several factors, including the dimensions (length, diameter) and stiffness of the object. However, we haven't said what happens *next*. In reality, an object can *really* be left to itself only in physics textbooks! If you push a swing once, it starts to oscillate, but the *amplitude* of its oscillations gradually decreases, and the swing will stop soon. This is due to the drag force exerted by air and to *friction* (see Chap. 14) at the level of junctions between the chains and the support structure. In general, the reduction of the motion of something which oscillates (an "oscillator") by an external force is called *damping*. When an object's harmonic motion is damped, the amplitude of the oscillations decreases exponentially over time, and their frequency is lower as compared with the (undamped) natural frequency. Of course, like any "real object," also invasive pressure monitoring systems undergo damped oscillations (and this is why we refer to the "exaggerated" oscillations due to resonance with the term "underdamping"). Hence, the "dynamic response" of these systems to the solicitation exerted by the pressure wave depends not only on their natural frequency, but also on the so-called *damping coefficient* (usually symbolized ζ), which is a measure of their "resistance" to oscillation or, in other words, of how rapidly they stop oscillating after the solicitation has ceased [2, 21]. The same factors affecting the natural frequency of a transducer-tubing-catheter system also affect (in the opposite direction) its damping coefficient ζ, according to an equation like this:

$$\zeta = \frac{4\mu}{r^3}\sqrt{\frac{d\,L}{\pi\,E}}$$

where r, L, d, and E are, respectively, radius, length, density, and volume elasticity (stiffness) (as in Eq. 19.33) [13, 16, 18]. When the damping coefficient is too high, the system oscillates less than it should to adequately "represent" the original arterial pressure waveform, leading to a "flat" curve with underestimated systolic pressure and overestimated diastolic pressure (overdamping). As is evident from the above equation, all the factors which reduce natural frequency (thus increasing the possibility of resonance and, hence, of underdamping) also increase the damping coefficient, possibly leading to overdamping. In particular, all factors which significantly reduce the volume elasticity E (stiffness) of the system by "absorbing" a lot of energy may lead to overdamping: these include air bubbles (usually larger than those causing resonance), blood clots, and the use of too soft tubing (this is why the common tube extensions used for intravenous infusions, which are usually much more compliant, i.e., less stiff, as compared with the tubes included in the invasive pressure monitoring kits, cannot be used for invasive pressure monitoring) [2, 13, 21]. Moreover, the lower its natural frequency, the narrower the range of damping coefficients within which the system is adequately damped, that is provide a reliable representation of the arterial pressure waveform [2]. In other words, the damping coefficient above which an invasive pressure monitoring system becomes overdamped is lower the lower is its natural frequency; accordingly, in addition to increase the likelihood of resonance (and underdamping), the reduction of natural frequency may also lead to overdamping (this is not only because the same factors which reduce the natural frequency also increase the damping coefficient, but also because a lower damping coefficient is sufficient to make the system overdamped if natural frequency is lower).

We all know that we can see with our own eyes the tendency of an invasive pressure monitoring system to resonate (underdamping) or its "reluctance" to oscillate (overdamping) with the simple *fast-flush test*, which even allows us to *calculate* both the natural frequency and the damping coefficient of the system [2, 13, 16, 21].

References

1. Szocik JF, Barker SJ, Tremper KK. Fundamental principles of monitoring instrumentation. In: Miller RD, editor. Miller's anesthesia. 6th ed. Philadelphia, PA: Elsevier Churchill Livingstone; 2005. p. 1191–225.
2. Mark JB, Slaughter TF. Cardiovascular monitoring. In: Miller RD, editor. Miller's anesthesia. 6th ed. Philadelphia, PA: Elsevier Churchill Livingstone; 2005. p. 1265–362.
3. Aston D, Rivers A, Dharmadasa A. Equipment in anaesthesia and critical care. Branbury: Scion Publishing; 2014. p. 375.
4. Pittman JA, Ping JS, Mark JB. Arterial and central venous pressure monitoring. Int Anesthesiol Clin. 2004;42(1):13–30.
5. Neice AE, Zornow MH. Xenon anesthesia for all, or only a select few? Anaesthesia. 2016;71(11):1267–72.
6. Law LS, Lo EA, Gan TJ. Xenon anesthesia: a systematic review and meta-analysis of randomized controlled trials. Anesth Analg. 2016;122(3):678–97.
7. Harris PD, Barnes R. The uses of helium and xenon in current clinical practice. Anaesthesia. 2008;63:284–93.
8. Jin Z, Piazza O, Ma D, Scarpati G, De Robertis E. Xenon anesthesia and beyond: pros and cons. Minerva Anestesiol. 2019;85(1):83–9.
9. Montgomery C. Why does inhaling helium make one's voice sound strange? Sci Am. 2004;291:122.
10. Salimbeni C. Phonation by means of various gaseous media. J Laryngol Otol. 1984;98:167–72.
11. Hess DR, Fink JB, Venkataraman ST, Kim IK, Myers TR, Tano BD. The history and physics of heliox. Respir Care. 2006;51:608–12.
12. Walker J. The flying circus of physics. 2nd ed. New York: Wiley; 2007. p. 64–5.

13. Mittnacht AJC, Reich DL, Sander M, Kaplan JA. Monitoring of the heart and vascular system. In: Kaplan's cardiac anesthesia: for cardiac and noncardiac surgery. 7th ed. The Netherlands: Elsevier; 2016. p. 390–426.
14. Aston D, Rivers A, Dharmadasa A. Equipment in anaesthesia and critical care. Branbury: Scion Publishing; 2014. p. 199–201.
15. Walker J. The flying circus of physics. 2nd ed. New York: Wiley; 2007. p. 126–7.
16. Gilbert M. Principles of pressure transducers, resonance, damping and frequency response. Anaesth Intensive Care Med. 2012;13(1):1–6.
17. Shapiro GG, Krovetz LJ. Damped and undamped frequency responses of underdamped catheter manometer systems. Am Heart J. 1970;80(2):226–36.
18. Kleinman B. Understanding natural frequency and damping and how they relate to the measurement of blood pressure. J Clin Monit. 1989;5(2):137–47.
19. Segers P, Stergiopulos N, Westerhof N. Relation of effective arterial elastance to arterial system properties. Am J Physiol Heart Circ Physiol. 2002;282(3):H1041–6.
20. Heimann PA, Murray WB. Construction and use of catheter-manometer systems. J Clin Monit. 1993;9:45–53.
21. Stoker MR. Principles of pressure transducers, resonance, damping and frequency response. Anaesth Intensive Care Med. 2004;5(11):371–5.

Ultrasounds and Doppler Effect: Echocardiography and Minimally Invasive Cardiac Output Monitoring

20

Contents

Where's Physics

Daily life	Ambulance sirens and Formula 1 races Speeding fines (speed radar gun) The echolocation system of bats The Universe is expanding
Physics involved	Sound waves (ultrasounds) Acoustic impedance Doppler Effect
Clinical practice	Ultrasonography/echocardiography Measuring blood velocity and cardiac output (CO) Esophageal Doppler cardiac output monitoring UltraSonic cardiac output monitor (USCOM) Transcranial Doppler

20.1 A Few Notes on Ultrasonography

In the last years, anesthesiologists and intensivists are relying more and more often, in their clinical practice, on *ultrasonography* (e.g., perioperative transesophageal echocardiography in both cardiac and noncardiac surgery [1–4]; ultrasound-guided

A. Pisano, *Physics for Anesthesiologists and Intensivists*, https://doi.org/10.1007/978-3-030-72047-6_20

peripheral nerve blocks [5, 6] or vascular access [7–9]; pleural and lung ultrasound [10–12]; and so on).

As mentioned in Chap. 19 (it is advisable to read at least Sect. 19.1 before proceeding with the reading of this chapter), *ultrasounds* are "sound" waves with a frequency above 20 kHz (i.e., 20,000 per second), which are not audible to the human ear. It is common knowledge that an ultrasound image is formed from *reflection* of ultrasound waves which occurs at boundaries between tissues with different *acoustic impedance* [13, 14]. The latter, usually symbolized Z, is roughly the "resistance" that a tissue opposes to the transmission of sound waves through it and is defined as the product of the tissue density d and the speed of sound v through it:

$$Z = d\,v \tag{20.1}$$

In particular, the higher the difference in the acoustic impedance between two neighboring tissues, the more intensely ultrasounds are reflected. In practice, since the speed of sound in soft tissues is relatively constant (about 1500 m/s, on average), reflection primarily occurs due to *density* changes.

The dependency of sound waves reflection on the difference in acoustic impedance at boundaries between different media is the reason why *ultrasound gel* (which has been already mentioned in Chap. 8 as an example of "pseudoplastic non-Newtonian fluid") is needed to get an ultrasound image. In fact, the acoustic impedance of air is much lower than that of our body, since both air density and the speed of sound through it (which, as said in Chap. 19, is around 340 m/s) are much lower as compared with any tissue. Accordingly, at the air/body boundary, almost all ultrasounds would be reflected before entering the body, unless you replace the air "entrapped" between the probe and skin with a substance (the ultrasound gel) with an acoustic impedance similar to that of tissues.

In diagnostic ultrasound, a "beam" of ultrasounds with frequency between 1 and 20 million Hertz (or 1–20 megahertz, symbolized MHz) is emitted by a set of *piezoelectric crystals,* which are made of a material which vibrates under the effect of an alternating electric current (see Chap. 11) and, conversely, undergoes a change in its electric polarization in response to an applied *pressure* (*piezoelectric effect*). Accordingly, the reflected ultrasound waves (which, as discussed in Chap. 19, are "pressure waves") are turned into an *oscillating electric signal* (and, then, into an image), thanks to the interaction with the same piezoelectric crystals which had previously produced them. The image is built according to the *time* which is needed for the waves to return to the probe; the waves reflected from more distant structures take more time to come back. This is exactly how bats identify obstacles and assess their distance (echolocation) [15].

Further "technological" details about the production and interpretation of ultrasound images are beyond the scope of this book and can be found in all ultrasound textbooks (such as [13]) as well as, nowadays, in many anesthesia and intensive care textbooks (e.g., [14]).

In the context of ultrasonography, the so-called *Doppler analysis* allows, as known, to measure blood velocity within blood vessels or cardiac chambers (as mentioned in Chap. 7). This technique relies on a behavior which is common to *all* traveling waves, known as *Doppler Effect*. The remaining part of this chapter is dedicated to the general description of this phenomenon, that we often encounter in daily life, as well as to some examples of its important applications in perioperative and intensive care medicine, such as echocardiography, minimally invasive cardiac output (CO) monitoring (esophageal Doppler), noninvasive ultrasonic CO monitoring, and transcranial Doppler.

20.2 Bats, Speeding Fines, Echocardiography, and Cardiac Output Monitoring: The Doppler Effect

If you have been ever overtaken by a police car or an ambulance with their sirens turned on, or if you have ever watched a Formula 1 race, then you should have heard in person the Doppler Effect. It consists in a change in the perceived *frequency* (or wavelength) of a wave whose source is in motion relative to the observer. More generally, this phenomenon occurs when either the wave source or the observer (or both) are moving relative to the medium through which the wave travels. In particular, *for a stationary observer and a moving source* (as in Fig. 20.1), the wave frequency *increases* if the source is approaching to the observer and *decreases* if it is moving away from it, according to the following equation:

$$f = f_0 \frac{v}{v \pm v_S} \tag{20.2}$$

where f is the frequency detected by the observer, f_0 is the frequency emitted by the source, v is the *wave speed* (e.g., through air), v_s is the speed of the source relative to the observer, and the sign ± is *negative* for an approaching source and *positive* for a receding one.

Of course, this is the same as saying that the wavelength *decreases* if the source is moving towards the observer and *increases* if it is moving away from it (see Fig. 20.1); in fact, as discussed in Chap. 16, the wavelength and the frequency of a wave which propagates with constant speed are inversely proportional. Accordingly, Eq. (20.2) (which provides the *frequency change* for a source that is moving with speed v_S relative to a stationary observer) can be easily derived from the *wavelength change* which is shown in the example of Fig. 20.1.

Consider the case of the source approaching the observer (Fig. 20.1a). If the segment *SS'* is the distance traveled by the source (with a constant speed v_S) over a time interval equal to the wave *period* T (see Chap. 16), then it is:

$$SS' = v_S T \tag{20.3}$$

In fact, as mentioned several times in this book (see Chapters 1, 7, 16, and Box 4.1), speed is "distance divided by time" and, hence, distance is "speed multiplied by time." Accordingly, the wavelength λ perceived by the observer will be:

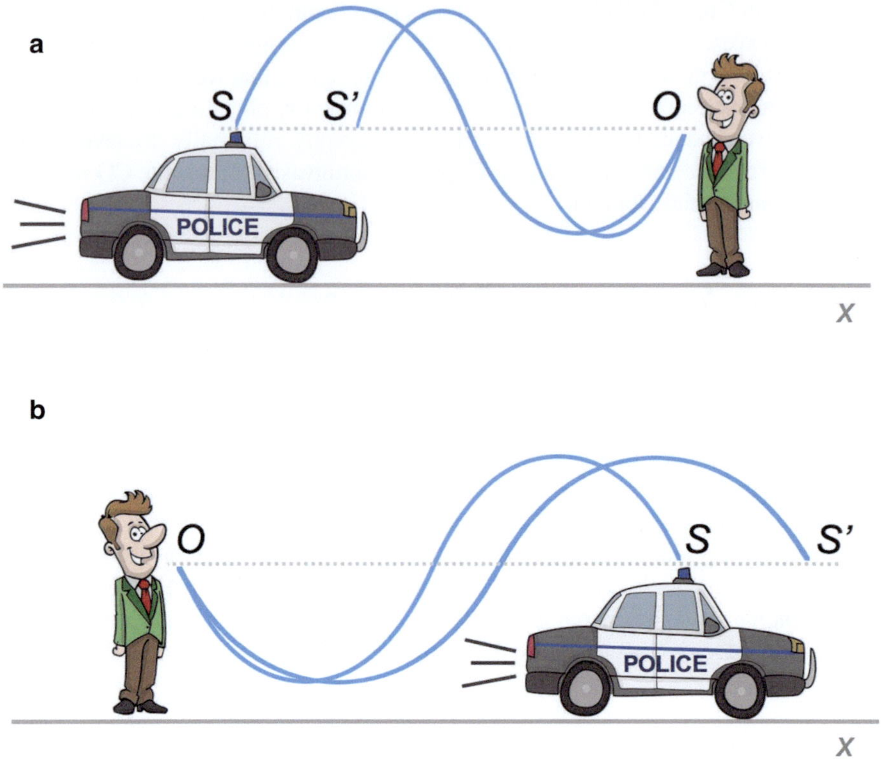

Fig. 20.1 Doppler Effect. (**a**) Source moving towards a stationary observer. Let *S* be the position of the sound source (i.e., the police siren) at the beginning of a complete oscillation, which ends (and then starts over again) at the point *O* where there is the observer. If the source were stationary, the wavelength of this wave would be the segment *OS* (note that the illustration represents a "snapshot" at a certain time instant as in Fig. 16.2b). However, if the source is moving so that it is at the point S′ after a time interval equal to the period of the wave (see Chap. 16), then the wavelength perceived by the observer will be *OS'*, i.e., it will be reduced by the length *SS'* (which is the space traveled by the source in a time interval equal to the wave period). Accordingly, the perceived frequency will be higher (since wavelength and frequency are inversely proportional). (**b**) Source moving away from a stationary observer; in this case, exactly the opposite occurs; hence, the perceived wavelength will be increased by the distance *SS'*, and the frequency will be lower

$$\lambda = OS - SS' = \lambda_0 - v_S T \tag{20.4}$$

with λ_0 (=OS) being the "original" wavelength (i.e., the wavelength that would be perceived by the observer if the source were stationary).

From Eq. (16.2) it is also:

$$T = \frac{\lambda_0}{v} \tag{20.5}$$

where v is the speed (of propagation) of the wave.

After substituting for T from Eq. (20.5), Eq. (20.4) becomes:

$$\lambda = \lambda_0 - v_S \frac{\lambda_0}{v} = \lambda_0 \left(1 - \frac{v_S}{v}\right) \tag{20.6}$$

Remembering that the speed v of a wave is the product of its frequency f and its wavelength (Eq. 16.3), and that the *wave* speed does not change whether the source is moving or not (the wave speed must not be confused with the source speed), it is:

$$v = f\lambda = f_0 \lambda_0 \tag{20.7}$$

or:

$$\lambda = \frac{v}{f} \tag{20.8}$$

and

$$\lambda_0 = \frac{v}{f_0} \tag{20.9}$$

where f_0 is the "original" frequency and f is the frequency as perceived by the observer. Substituting from Eqs. (20.8) and (20.9), Eq. (20.6) can be written as:

$$\frac{v}{f} = \frac{v}{f_0}\left(1 - \frac{v_S}{v}\right) \tag{20.10}$$

or, noting that the "v" on the left and that on the right (outside the parenthesis) cancel each other out:

$$f = f_0 \frac{1}{1 - \frac{v_S}{v}} \tag{20.11}$$

After multiplying the numerator and denominator by v, Eq. (20.11) becomes:

$$f = f_0 \frac{v}{v - v_S} \tag{20.12}$$

which is none other than Eq. (20.2) in the case of a source approaching the observer. It is easy to imagine that if we do it all over again, but this time considering the *increase* in wavelength (from *OS* to *OS'* = *OS* + *SS'*) which occurs when the source moves away from the observer (Fig. 20.1b), the result would be:

$$f = f_0 \frac{v}{v + v_S} \tag{20.13}$$

and with this we have completed the proof of Eq. (20.2) (Doppler Effect).

Since, as discussed above, a lower (fundamental) frequency of a sound wave means a less acute sound, the pitch of an ambulance siren which is receding from you is characteristically "waning." This effect is particularly evident during a Formula 1 race, due to the high speed of the wave source (i.e., the cars); the typical "wroom" made by the cars whizzing in front of you is rising (in its pitch) during "wro" and waning during "om."

As mentioned, Doppler Effect does not concern only sound waves but also electromagnetic waves. For example, the wavelength of light coming from a source which is moving away from an observer increases. Since the longer wavelengths (within the visible range) are those of red, it is said that light emitted by receding objects is "red shifted," i.e., shifted *towards* (but not necessarily to) the red. The observation of the *red shift* of light coming from galaxies led to the conclusion that the Universe is expanding, and is one of the proofs of the so-called Big Bang theory (which is nowadays considered "a fact" rather than a theory). Similarly, the police can get the proof of your speeding, thanks to the Doppler Effect; a "speed radar gun," one of the several methods for speed detection, uses *radio waves* (see Chap. 16) "fired" against a moving car to measure its speed according to the variation in wavelength (which is often referred to as *Doppler shift*) of the reflected waves.

Also bats rely on the Doppler shift of ultrasounds they emit in order to estimate their own speed and to distinguish moving objects (e.g., insects) [15]. As discussed below, many echocardiographic measures, as well as some techniques commonly used for perioperative monitoring (such as esophageal Doppler, ultrasonic noninvasive CO monitoring, and transcranial Doppler), use the Doppler Effect in a similar way. In practice, a beam of ultrasounds of known frequency f_0 is emitted towards the flowing blood and travels through it with a known speed v, which is the speed at which ultrasounds (and sound waves in general) propagate through blood. This beam is reflected by blood and comes back towards the probe with the same speed v, but *with a different frequency* f, since the reflected wave comes from a moving source (i.e., the flowing blood). In particular, frequency increases if blood flow is directed towards the probe, and decreases if blood flows in the opposite direction. The frequency of the reflected ultrasounds can be measured by means of the piezoelectric effect (see above), and then the source speed v_S, which is none other than blood flow velocity, remains the only unknown in Eq. (20.2) and can be thus calculated according to it.

20.2.1 Cardiac Doppler Ultrasound

In echocardiography textbooks, Eq. (20.2) is usually written as follows (and referred to as the *Doppler equation*) [13, 14]:

$$v_{\text{blood}} = \frac{v\left(f_{\text{r}} - f_{\text{e}}\right)}{2 f_{\text{e}}\left(\cos\theta\right)} = v\,\Delta f / 2 f_e \cos\theta \qquad (20.14)$$

where v_{blood} is the blood flow velocity; v is the speed of sound in blood (1540 m/s); Δf is the Doppler shift, i.e., the difference between the frequency emitted by the probe (f_e) and that reflected to it (f_r); 2 accounts for the double travel of the ultrasound beam (from the probe to the reflecting source and then back to the probe); and θ is the intercept angle between the blood flow direction and the ultrasound beam.

The term cosθ "appears" since velocity is a vector (see Chap. 13); evidently, it is the component of the blood flow velocity parallel to the direction of the ultrasound beam which causes the Doppler shift. Accordingly, a close alignment between the ultrasound beam and the direction of blood flow is pivotal in cardiac Doppler measurements ($\theta < 20°$ is usually acceptable), while no Doppler shift can be recorded at 90° (in fact, as mentioned in Chap. 1, cos 90° = 0) [13].

Two echocardiographic applications of Doppler analysis have been already discussed in Chap. 7: the measurement of blood flow velocity at both the aortic valve and the left ventricle outflow tract (LVOT) level allows to estimate, according to the continuity equation, the "actual" aortic valve area (AVA), while the pressure gradients across heart valves can be calculated from blood flow velocity, thanks to the simplified Bernoulli's equation. Both techniques are used, for example, for the grading of valve stenosis.

Another example is the estimation of systolic pulmonary artery pressure as the sum of right atrial pressure and the systolic pressure gradient across the tricuspid valve (calculated from the maximum regurgitant jet velocity, which in turn is measured according to Eq. (20.14)) [13].

Let's see, finally, how cardiac output (CO) can be measured. As known, CO is the product of heart rate and stroke volume (SV), that is the volume of blood ejected from the ventricle at each systole. Left ventricle SV can be calculated as the volume of blood ejected into the aorta at each systole, which can be approximated to the volume of a cylinder whose base area is the systolic cross-sectional area (CSA) of the left ventricle outflow tract and whose height is the mean distance traveled by blood cells during a systole. Since, as discussed in detail in Chap. 1 (Sect. 1.4.2), velocity is the derivative of distance (or, if we want to be picky, of displacement) over time, the integral of velocity over time is distance (in fact, as we said, integration is the inverse of differentiation). Accordingly, the mean distance traveled by blood cells during a systole can be calculated as the integral of the Doppler systolic velocity-time curve, which is called velocity-time integral (*VTI*). Remembering that the volume of a cylinder is the product of its base area and its height, it is:

$$\mathrm{SV} = \mathrm{CSA} \times \mathrm{VTI} \tag{20.15}$$

The CSA is usually estimated by measuring (sonographically) the diameter of LVOT either at the aortic valve leaflet tips or in the ascending aorta, while the VTI is obtained from the measurement of aortic flow velocity by Doppler analysis at the same level [13].

20.2.2 Doppler-Based Minimally Invasive or Noninvasive Monitoring Devices

In Chap. 10, we discussed how the laws of thermodynamics allow the measurement of cardiac output (CO) by means of a pulmonary artery catheter (PAC) and the thermodilution method. Although this method still remains the gold standard for CO assessment, minimally invasive devices (e.g., pulse contour analysis monitors and esophageal Doppler monitors) or noninvasive devices (e.g., transthoracic electrical bioimpedance monitors) are available for CO monitoring in several clinical settings [16]. Some of these monitors rely on Doppler analysis.

Esophageal Doppler CO monitors are commonly used for hemodynamic monitoring (for example to guide fluid therapy) in both the operating room and the intensive care unit [17, 18]. The way these devices measure the stroke volume is very similar to the echocardiographic method described above: the blood flow velocity in the *descending aorta* is measured, according to Eq. (20.14), by means of a Doppler ultrasound probe placed in the esophagus (for this reason, these monitoring systems are referred to as "minimally invasive"). The curve of flow velocity over time is then integrated to obtain the so-called *stroke distance* (i.e., the distance traveled by blood at each systole; in practice, the analogous of the "echocardiographic" *VTI*). The stroke distance, multiplied by the cross-sectional area (CSA) of the descending aorta, provides the *descending aortic stroke volume* (which is about 70% of the stroke volume). However, the most commonly used esophageal Doppler CO monitor (CardioQ-ODM, Deltex Medical, Chichester, UK) [19] neither measures nor estimates the CSA. Conversely, it translates the stroke distance into an absolute value of stroke volume using an algorithm developed by comparison with a thermodilution technique (in practice, it uses a nomogram based on patient's age, weight, and height to convert the stroke distance measured in a certain patient in a value of stroke volume).

The ultrasonic cardiac output monitor (USCOM) (Uscom Ltd., Sidney, Australia) also relies on ultrasounds and Doppler analysis to measure the VTI according to the blood velocity profile across either the aortic valve or the pulmonary valve and on a nomogram to estimate valve area, but it is totally noninvasive since it uses a cutaneous probe placed at the sternal or left hemithoracic level (to insonate the aortic valve or the pulmonary valve, respectively) [16, 20].

Of course, both these devices have their best usefulness as trend monitors.

Finally, *Transcranial Doppler* (TCD) is another monitoring tool which is based on the Doppler equation. It uses a beam of ultrasounds to measure blood flow velocity within cerebral vessels such as the middle cerebral artery. TCD is useful, for example, for detection of cerebral emboli and as a trend monitoring of cerebral blood flow during cardiac surgery [21].

References

1. Mahmood F, Shernan SK. Perioperative transoesophageal echocardiography: current status and future directions. Heart. 2016;102(15):1159–67.
2. Reeves ST, Finley AC, Skubas NJ, Swaminathan M, Whitley WS, Glas KE, et al. Council on Perioperative Echocardiography of the American Society of Echocardiography and the Society of Cardiovascular Anesthesiologists. Special article: basic perioperative transesophageal echocardiography examination: a consensus statement of the American Society

of Echocardiography and the Society of Cardiovascular Anesthesiologists. Anesth Analg. 2013;117(3):543–58.
3. Vegas A. Perioperative transesophageal echocardiography in cardiac surgery procedures. In: Cheng DC, Martin J, David T, editors. Evidence-based practice in perioperative cardiac anesthesia and surgery. Cham: Springer; 2021. p. 369–80.
4. Fayad A, Shillcutt SK. Perioperative transesophageal echocardiography for non-cardiac surgery. Can J Anaesth. 2018;65(4):381–98.
5. Lewis SR, Price A, Walker KJ, McGrattan K, Smith AF. Ultrasound guidance for upper and lower limb blocks. Cochrane Database Syst Rev. 2015 Sep 11;9:CD006459.
6. Merella F, Mossetti V. Ultrasound-guided upper and lower extremity nerve blocks in children. BJA Education. 2020;20(2):42–50.
7. Lamperti M, Biasucci DG, Pittiruti M, et al. European Society of Anaesthesiology guidelines on peri-operative use of ultrasound for vascular access (PERSEUS vascular access). Eur J Anaesthesiol. 2020;37:344–76.
8. Reusz G, Csomos A. The role of ultrasound guidance for vascular access. Curr Opin Anaesthesiol. 2015;28(6):710–6.
9. Weiner MM, Geldard P, Mittnacht AJ. Ultrasound-guided vascular access: a comprehensive review. J Cardiothorac Vasc Anesth. 2013;27(2):345–60.
10. Vignon P, Repessé X, Vieillard-Baron A, Maury E. Critical care ultrasonography in acute respiratory failure. Crit Care. 2016;20(1):228.
11. Lichtenstein D. Lung ultrasound in the critically ill. Curr Opin Crit Care. 2014;20(3):315–22.
12. Piette E, Daoust R, Denault A. Basic concepts in the use of thoracic and lung ultrasound. Curr Opin Anaesthesiol. 2013;26(1):20–30.
13. Otto CM. Textbook of clinical echocardiography. 6th ed. The Netherlands: Elsevier; 2018.
14. Kahn RA, Maus T, Salgo I, Weiner MM, Shernan SK. Basic intraoperative transesophageal echocardiography. In: Kaplan's cardiac anesthesia: for cardiac and noncardiac surgery. 7th ed. The Netherlands: Elsevier; 2016. p. 427–504.
15. Walker J. The flying circus of physics. 2nd ed. New York: Wiley; 2007. p. 156.
16. Drummond KE, Murphy E. Minimally invasive cardiac output monitors. Cont Educ Anaesth Crit Care Pain. 2012;12(1):5–10.
17. Schober P, Loer SA, Schwarte LA. Perioperative hemodynamic monitoring with transesophageal Doppler technology. Anesth Analg. 2009;109(2):340–53.
18. Aston D, Rivers A, Dharmadasa A. Equipment in anaesthesia and critical care. Branbury: Scion Publishing; 2014. p. 276–9.
19. Vincent JL, Pelosi P, Pearse R, Payen D, Perel A, Hoeft A, et al. Perioperative cardiovascular monitoring of high-risk patients: a consensus of 12. Crit Care. 2015;19:224.
20. Elgendy A, Seppelt IM, Lane AS. Comparison of continuous-wave Doppler ultrasound monitor and echocardiography to assess cardiac output in intensive care patients. Crit Care Resusc. 2017;19(3):222–9.
21. Edmonds HL, Gordon EK, Levy WJ. Central nervous system monitoring. In: Kaplan's cardiac anesthesia: for cardiac and noncardiac surgery. 7th ed. The Netherlands: Elsevier; 2016. p. 677–97.

Part IX

And Finally...

Activated Clotting Time and A Brief Look at Relativity

21

Contents

Where's Physics

Daily life	So works the world (really!) GPS navigation systems
Physics involved	Special Relativity (Time dilation)
Clinical practice	Getting the result of Activated Clotting Time faster (just for fun)

21.1 Surgeons are Always in a Hurry

Activated clotting time (*ACT*), that is, the time (in seconds) needed for clot formation in a sample of whole blood after the addition of an activator, is still the most common method used to monitor the effectiveness of heparin anticoagulation in cardiac surgery [1]. Since an *ACT* above 480 s is usually required for cardiopulmonary bypass (CPB), it takes at least 8 min (480/60) from the time of sampling to get a useful result. Moreover, although the peak *ACT* prolongation has been reported to occur in less than one minute, it is usually advisable to wait 2–5 min after heparin administration before taking a blood sample for *ACT* [2]. Accordingly, even making a little discount (i.e., allowing surgeons to start CPB with an *ACT* above 400 s), it takes about 10 min from heparin administration to the onset of CPB.

A. Pisano, *Physics for Anesthesiologists and Intensivists*,
https://doi.org/10.1007/978-3-030-72047-6_21

These 10 min, however, often seem to last an eternity, with surgeons who repeatedly ask: "which is the *ACT*?", "can I go on CPB?". It would be very nice to be able, say, to halve this time interval.

As discussed below, this is indeed *possible*… According to (Einstein's) *Relativity*! [3–8].

21.2 How to Get the Result of *ACT* Faster

One of the principles of Relativity is *time dilation*: for someone or something who is moving, *time flows slower* (as compared with the reference frame in which that person/object is moving). In other words, *a clock ticks slowly when it moves* [3]. If, after the administration of heparin, the surgeon went out for a run around the hospital (*at an appropriate speed*), (s)he might find the *ACT* ready in 5 min! In fact, for the surgeon, time would flow slower than for those who remained in the operating room (i.e., the reference frame with respect to which the surgeon runs).

Before proceeding with the quantitative analysis of this phenomenon (and with its proof!), let's see the two fundamental ideas on which Relativity relies.

21.2.1 The "Relativity" and the "Speed of Light" Postulates

Relativity is based on two basic postulates:

1. *All* the laws of physics apply in the same way whatever the *inertial reference frame* from which they are observed (*relativity postulate*). The term "inertial" refers to the fact that the *law of inertia,* or *Newton's first law* (see Chap. 14), must be valid. In other words, inertial reference frames are those reference frames in which a body moves *with constant speed* (or is motionless), which is the same as saying that inertial reference frames move relative to the body (and to each other) with constant speed. For example, if you walk with a friend at a speed of 3 km/h (in the travel direction) along a railroad car which is entering the station at a speed of 40 km/h, you are motionless relative to your friend, while you are moving at 3 km/h relative to the car and at 43 (i.e., 40 + 3) km/h relative to the station (this is a good approximation at these low speeds although velocities do not add up in this way according to Einstein's relativity). Your friend, the railroad car, and the station are *all* inertial reference frames (for you) which move relative to each other with different constant speeds. The part of Relativity which deals with inertial reference frames is called *special relativity* (which includes the principle of time dilation further discussed below), while *general relativity* describes the effects of acceleration (and, hence, of *gravity*) on space-time and matter.
2. The *speed of light c* (in vacuum) is the same in all inertial reference frames (*speed of light postulate*). This is a fundamental point; it means that, if you walk at a speed of 3 km/h along the above-mentioned railroad car, holding a flashlight (turned on) in your hand, the speed of its light is not c relative to you, $c + 3$ km/h

for someone sitting in the railroad car, and c + 43 km/h for the stationmaster (who is motionless relative to the station), but is exactly c for *all* three.

21.2.2 How Fast Should the Cardiac Surgeon Run

Figure 21.1 shows a classical representation of the *light clock*, which is used in all physics textbooks to prove the Einstein's equation for *time dilation*. This equation will tell us to what extent time will slow for the surgeon, depending on his/her speed, as compared to the inertial reference frame in which (s)he moves (e.g., the operating room).

A light clock consists in a light source and a light detector surmounted by a mirror at a distance h. The light source emits a pulse of light towards the mirror, where it is reflected and comes back to the detector (we discussed the *law of reflection* in Chap. 18). The clock "ticks," i.e., it advances by a time interval Δt_0, whenever the detector receives a light pulse, leading, at the same time, to the emission of another light pulse from the source. Hence, when the clock is observed from its own reference frame, every "tick" corresponds to a time interval Δt_0. During this time interval, light travels, approximately, a distance 2 h (assuming that the light source and the light detector are very close).

Now imagine that the light clock is moving at a constant speed v (e.g., we have put it in our checked luggage during a flight). Figure 21.2 shows what happens when one observes the clock from the reference frame in which the clock moves (therefore, not from the airplane but, for example, from the airport). As above, one "tick" of the clock corresponds to the time interval Δt which is needed for light to travel from the source to the mirror and then to the detector. In the same time interval, the

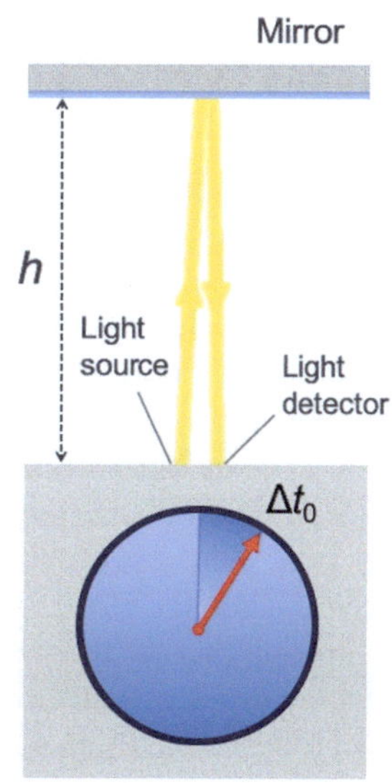

Fig. 21.1 The "light clock" (see text)

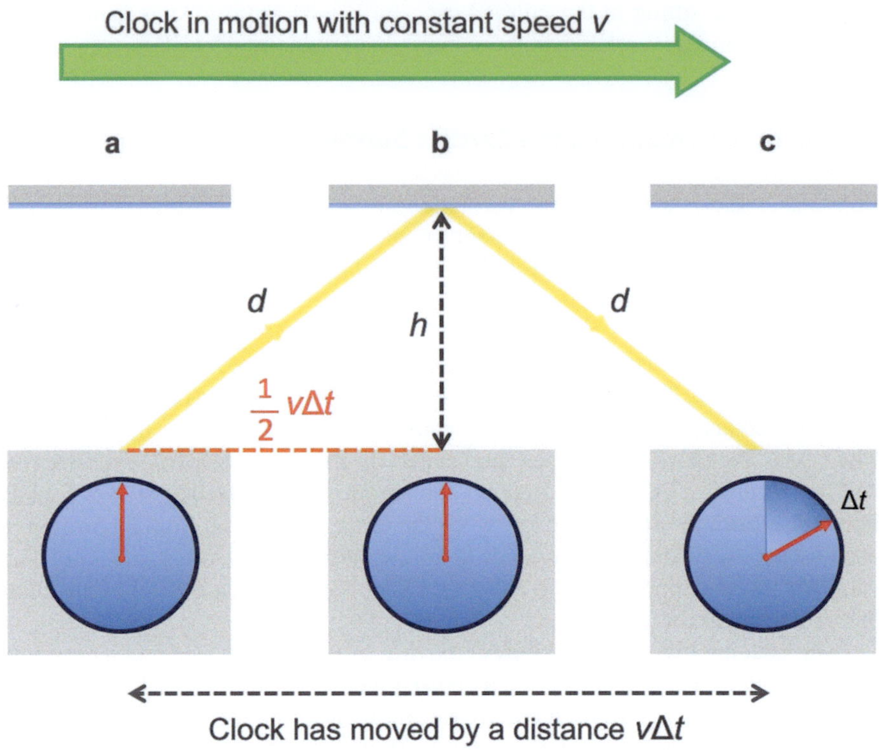

Fig. 21.2 A light clock observed from the inertial reference frame in which it moves with constant speed v. The three drawings (**a**), (**b**), and (**c**) represent three different positions in the space which the clock occupies during the time interval Δt

clock moves by a distance $v\Delta t$. In fact, as recalled many times in this book (see Chaps. 1, 7, 16, 20, and Box 4.1), if a body moves with a *constant* speed v, this speed is equal to the distance Δx which the body travels during the time Δt:

$$v = \frac{\Delta x}{\Delta t} \tag{21.1}$$

Accordingly, it is $\Delta x = v\Delta t$.

Evidently, during the time needed for the light to travel from the source to the mirror, the clock moves half of this distance, i.e., $v\Delta t/2$.

It appears clear from Fig. 21.2 that, when the clock is observed from the reference frame in which it moves, light covers a distance d from the source to the mirror (or from the mirror to the detector) which is longer than the distance h which light travels when the clock is observed in its own reference frame (in which it is at rest).

Noting that the three distances d, h, and $v\Delta t/2$ form, respectively, the hypotenuse and the other two sides of a right triangle, the distance d can be easily calculated according to the Pythagorean theorem (see Chap. 1):

$$d^2 = h^2 + \left(\frac{1}{2} v\Delta t\right)^2 \tag{21.2}$$

or

$$d = \sqrt{h^2 + \left(\frac{1}{2} v\Delta t\right)^2} \tag{21.3}$$

Hence, during the time interval Δt (i.e., between two "ticks"), light travels a distance:

$$2d = 2\sqrt{h^2 + \left(\frac{1}{2} v\Delta t\right)^2} \tag{21.4}$$

According to Eq. (21.1), and remembering that the speed of light is c, it is:

$$c = \frac{2d}{\Delta t} \tag{21.5}$$

and, therefore:

$$2d = c\,\Delta t \tag{21.6}$$

Putting Eqs. (21.4) and (21.6) together, we have:

$$2\sqrt{h^2 + \left(\frac{1}{2} v\Delta t\right)^2} = c\,\Delta t \tag{21.7}$$

On the other hand, in the clock's own reference frame (i.e., that of the airplane), light travels, as stated, the distance 2 h in the time interval Δt_0. According again to Eq. (21.1), it is:

$$2h = c\,\Delta t_0 \tag{21.8}$$

or

$$h = \frac{1}{2} c\Delta t_0 \tag{21.9}$$

After substituting for h from Eq. (21.9), Eq. (21.7) becomes:

$$2\sqrt{\left(\frac{1}{2} c\Delta t_0\right)^2 + \left(\frac{1}{2} v\Delta t\right)^2} = c\,\Delta t \tag{21.10}$$

and, after some algebra:

$$\Delta t = \frac{\Delta t_0}{\sqrt{1-(v/c)^2}} \tag{21.11}$$

Equation (21.11) is the equation of special relativity for time dilation, i.e., the quantitative relationship between the time Δt (which corresponds to the time interval between two clock "ticks" in a frame reference in which the clock moves with speed v) and the time Δt_0 (i.e., the time interval between two clock "ticks" in the clock's own reference frame) [3]. This means that, for those who are on the airplane, *life* itself flows more slowly as compared with those who are at the airport. Therefore, if there were two twins, one at the airport and one on the airplane, the traveler, on his/her return, would be younger than the brother remained at the airport (this is the first part of the famous *twin paradox*).

Let's now calculate the speed at which the surgeon should travel in order to halve his/her waiting for the *ACT*, i.e., the speed for which it is $\Delta t = 2\Delta t_0$ (in other words, the speed for which one minute in the surgeon's own reference frame lasts 2 min in that of the operating room). In order to do this, we just need to replace Δt with $2\Delta t_0$ in Eq. (21.11):

$$2\Delta t_0 = \frac{\Delta t_0}{\sqrt{1-(v/c)^2}} \tag{21.12}$$

Accordingly, Δt_0 is cleared, and the equation becomes:

$$\sqrt{1-(v/c)^2} = \frac{1}{2} \quad \text{or} \quad 1-(v/c)^2 = \frac{1}{4} \tag{21.13}$$

and, after a few simple steps:

$$v = \frac{\sqrt{3}}{2}c \cong 0.87c \tag{21.14}$$

Remembering that, as mentioned in Chaps. 14, 16, and 18, the speed of light (in vacuum) c is 300,000 kilometers/second (or 186,000 miles/second), i.e., *one billion kilometers/hour* (or *670 million miles/hour*), the surgeon should run at a speed of 870 million kilometers/hour (or about 583 million miles/hour), that is, about 700,000 times higher than that of a supersonic jet.

As you can easily realize, surgeons had better resign themselves to wait 10 min!

21.3 After All, It was Just for Fun

The fact that the effects of Relativity become evident only for incredibly high speeds, which probably we will never reach, does not mean that Relativity doesn't concern our everyday life. For example, Global Positioning System (GPS) navigation devices would not be so precise without considering the effects of Relativity on the satellites' clocks (which include both the "special relativity" effects, related to the high speed of satellites, and the "general relativity" effects, related to gravity) [3]. Anyhow, it is quite incorrect (and limiting) to talk only about a single aspect of Relativity, which should always be considered as a (beautiful and complex) whole.

In addition to time dilation, indeed, other relativistic phenomena which occur in moving objects (such as *length contraction*, *mass increase*, and the so-called *relativity of simultaneity* [3–8]), taken together, give a sense to some apparent paradoxes. However, the aim of this final chapter was just to have some fun with (harder) physics taking inspiration from an "anesthesiological" scenario (it is evident that the *ACT* was just a pretext). This is also to remember that, today, the term "physics" includes much more than what was dealt with in the previous 19 chapters of this book. In fact, although we have sometimes fleetingly peeked into the world of *quantum mechanics* (e.g., in Boxes 15.1 and 16.1), this book deals almost entirely with topics of *classical physics* (i.e., what comes before Relativity and Quantum Physics), in particular those with interesting applications in the field of anesthesia and intensive care medicine. The most curious of you will appreciate (I hope) the hints contained in this chapter (together with those about quantum physics in Chaps. 14–16) to broaden their horizons about physics.

References

1. Shore-Lesserson L, Enriquez LJ, Weitzel N. Coagulation monitoring. In: Kaplan's Cardiac Anesthesia: for cardiac and noncardiac surgery. 7th ed. The Netherlands: Elsevier; 2016. p. 698–727.
2. Shore-Lesserson L, Finley A, Murphy GS, Gravlee GP. Anticoagulation for cardiopulmonary bypass. In: Gravlee GP, Davis RF, Hammon JW, Kussman BD, editors. Cardiopulmonary bypass and mechanical support: principles and practice. 4th ed. Philadelphia, PA: Wolters Kluwer; 2016. p. 95–108.
3. Styer DF. Relativity for the questioning mind. Baltimore, MD: The Johns Hopkins University Press; 2011.
4. Durell CV. Readable relativity. New York: Dover; 2003.
5. Einstein A. Relativity: the special & the general theory. 100th Anniversary Edition. Princeton University Press: Princeton, NJ; 2015.
6. Gribbin J. Einstein's masterwork: 1915 and the general theory of relativity. London: Icon Books; 2015.
7. Fischer K. Relativity for everyone. How space-time bends. Cham: Springer; 2015.
8. Morin DJ. Special relativity: for the enthusiastic beginner. Scotts Valley: CreateSpace Independent Publishing Platform; 2017.

Index

A. Pisano, *Physics for Anesthesiologists and Intensivists*, https://doi.org/10.1007/978-3-030-72047-6

FSC
www.fsc.org
MIX
Papier aus verantwortungsvollen Quellen
Paper from responsible sources
FSC® C105338

If you have any concerns about our products,
you can contact us on
ProductSafety@springernature.com

In case Publisher is established outside the EU,
the EU authorized representative is:
Springer Nature Customer Service Center GmbH
Europaplatz 3, 69115 Heidelberg, Germany

Printed by Libri Plureos GmbH
in Hamburg, Germany